DATE DUE

DEC 22 1987

D0407728

Automatic Radar Plotting Aids Manual

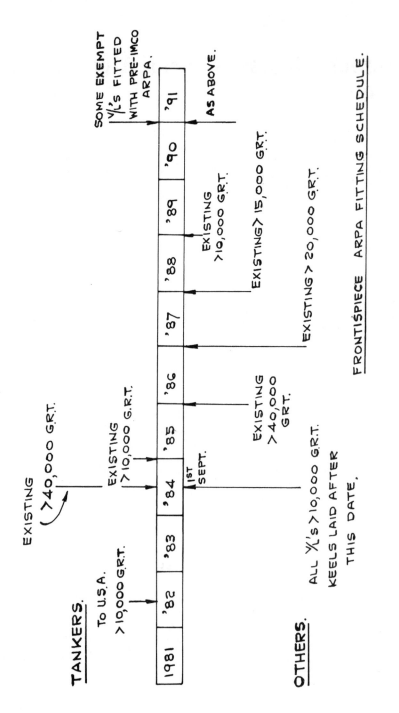

FRONTISPIECE   ARPA FITTING SCHEDULE.

ARPA fitting schedule

Automatic Radar Plotting Aids Manual

A Mariner's guide to the Use of ARPA

A. G. BOLE
Extra-Master Mariner, FRIN, FNI

K. D. JONES
Extra-Master Mariner, MPhil, FRIN, FNI

Cornell Maritime Press
Centreville, Maryland

ISBN 0-87033-285-6
Library of Congress Catalog Card Number: 81-71212

Manufactured in Great Britain
First American edition, 1982

Preface

The carriage of ARPA systems becomes mandatory for tankers (carrying hazardous cargoes) of more than 10,000 tons trading to the USA from July 1982, and it becomes progressively mandatory worldwide for other vessels of more than 10,000 tons from September 1984. Certification, following approved training for all Masters, Chief Mates and Officers in charge of a navigational watch on ships carrying ARPA, will probably be required.

With the fitting of ARPA on ships thus increasing, and the requirement for officer training by IMCO becoming more urgent, it was felt that there was an international need for a manual of information to supplement the period of training spent on ARPA Simulator courses. Practical exercises will naturally predominate in these courses but since present-day ARPA is based on new and modern technology, a much wider appreciation of the underlying principles and techniques of operation is necessary if the full potential of the installation is to be realized.

The manual follows IMCO specifications and is intended both for maritime students and for navigating and watchkeeping officers at all levels. Its use by students on ARPA courses will reduce their need for note-taking, allowing them to spend the maximum amount of time on practical exercises while making available to them the theory on which the practice must be based. For mariners, the manual provides material not normally covered in manufacturers' handbooks and is intended as a reference book for use at sea.

It is based on the research and experience gained by the authors at sea and on the Liverpool Polytechnic ARPA Simulator.

A. G. Bole
K. D. Jones

Acknowledgements

We would like to thank all the ARPA manufacturers for their help − ashore and afloat − and helpful suggestions both for the course and the manual, and Liverpool Polytechnic, in whose laboratories the lessons passed on in this manual were researched and learned. In particular we would like to thank Mr A. Tuthill of Racal Decca for his careful reading of the final manuscript and for his good, sound advice, also Captain J. P. O'Sullivan of Sperry Marine for his helpful and constructive comments.

We would especially like to thank Capt. W. O. Dineley for his contribution on the section relating to the relevance of the Collision Regulations and Dr. C. B. Barrass for his work on the captioning of the illustrations. For use of the radar screen illustration incorporated in the front cover design, permission from Sperry Marine is gratefully acknowledged.

Contents

Contents

Introduction

Modern technology has made it possible to present to an observer an analysis of the movement of any ships in the vicinity of his own ship. Those systems which actually extract data from raw radar information and present it in any form, without the observer being actively involved in the tracking, plotting or analysis of the data, are classified as Automatic Radar Plotting Aids. Other equipments which require the observer to do some of the plotting himself, even if this only involves the movement of preselected lines on the display, are usually classed as Assessment or Appraisal Aids and do not constitute an approved ARPA.

What constitutes an approved ARPA?

In recent years, many aids to radar plotting have appeared, but only those which conform to the IMCO, ARPA Specification (see Appendix I) are likely to be acceptable where there is a legal requirement to fit ARPA. The IMCO Specification will form the basis of a future national legislation; the national specification may differ in detail from the IMCO Specification, but it will in no way be inferior to it. Manufacturers are now indicating which of their equipments conform with the IMCO and other national specifications.

Who must fit ARPA?

This is quoted in full later, but summarized here for quick reference:

1 All ships of 10,000 grt or more – keels laid after 1st September 1984.
2 Existing tankers 40,000 grt or more . . . from 1st January 1985.

3 Existing tankers 10,000 grt or more . . . from 1st January 1986.
4 Existing ships 40,000 grt or more . . . from 1st September 1986.
5 Existing ships 20,000 grt or more . . . from 1st September 1987.
6 Existing ships 15,000 grt or more . . . from 1st September 1988.

Two slight complications exist as a result of unilateral action by the US Government concerning vessels trading to the United States, and areas under their jurisdiction:

1 Vessels of 10,000 grt or more which carry oil or hazardous material will be required to be fitted with an approved ARPA by 1st July 1982.
2 Vessels fitted with an ARPA to the earlier MARAD specification need not conform with IMCO Specification until 1st January 1991. (*Note*: this may also apply in some other countries).

Note: It would appear that already (1981) there is a call to remove the tonnage limit on vessels carrying hazardous materials.

Who will have to hold a certificate to operate ARPA?

It is anticipated that the Master and *all* watchkeeping officers on a vessel equipped with an ARPA whether or not the ship is required by law to carry that ARPA will have to be certificated. Special ARPA courses, conducted on ARPA simulators, covering the elements considered in this manual will be available and satisfactory completion of such a course will entitle the candidate to a certificate. The normal entry requirement to these courses in the United Kingdom will be the possession of a Radar Simulator Course certificate as approved by the DoT.

Exact detail of both ARPA equipment and training course requirement appear in Appendices I and II.

Chapter 1

The Basic Radar System

1.1 What is a radar system?

A simple block schematic diagram of a radar is shown in Figure 1.1. This shows that there are four main elements.

(a) *A transmitter*; which generates pulses of electromagnetic energy at either 3 or 10 cm wavelength, and feeds them via a waveguide to an aerial. The number of these pulses transmitted in one second of time is termed the PRF and is typically in the region of 1,000. The length of the burst of energy in each pulse is termed the pulse length and is measured in time. Pulse lengths are typically less than one millionth part of a second (i.e. 1 μsec).

(b) *An antenna*; which shapes the energy and beams it out to space. The beam is generally considered to be fan shaped, being narrow in the horizontal plane but wider in the vertical plane. The same antenna gathers the energy, reflected from a target, when it arrives back at the ship. The beam rotates continuously in azimuth.

(c) *A receiver*; which handles the returned pulse from the aerial, amplifying, shaping and processing the response until it is fed to the display.

(d) *A display*; which is usually a long persistence phosphor, cathode ray tube. On this tubeface the echo of the target is displayed as a bright spot at a distance from the origin of the display which is proportional to the actual range of the target.

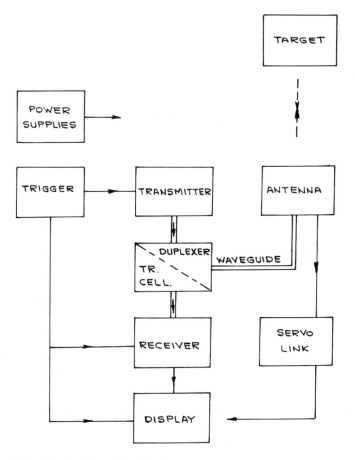

Figure 1.1 Block schematic of radar system

1.2   What governs the echo presentation?

The strength or brightness, the dimensions and positional accuracy of this displayed spot, or echo, depends on several factors. Some of these factors are inherent in the radar display, others are due to the media through which the pulse travels and the reflective performance of the target which is eventually struck by the pulse of energy. The factors may be listed as: −

(a) *The characteristics of the transmitter* such as:
  • the frequency of the transmitted energy
  • the number of pulses transmitted per second (PRF)

- the peak power of the pulse
- the length of the pulse

all govern the amount of energy which leaves the wave-guide on its way to the target.

(b) *The characteristics of the antenna* such as:
- the polarization of the radiated pulse
- the width of the aperture
- the type of the aperture
- its height above the sea surface

govern the shape of the beam. A wider aperture will concentrate the energy density. The siting of the antenna affects the arcs over which the energy can be radiated and hence detect targets. Further, the speed of rotation will govern the length of time for which a target lies in the beam.

(c) *The chacteristics of the target* such as:
- attitude, size and shape
- material and construction
- distance from the radar antenna
- height above the sea surface

(d) *The characteristics of the media through or over which the energy passes* such as:
- sea and waves
- rain and other precipitation

(e) *The characteristics of the receiver* such as:
- the type of amplifier and its sensitivity
- the bandwidth of the receiver
- the amount of 'cleaning' applied to the echo train
- limiting or differentiation applied to the video signal pulse shaping

(f) *The characteristics of the display* such as:
- the size of the smallest spot obtainable on the phosphor
- the sensitivity and persistence of the phosphor
- the dynamic range of the display

(g) *Additional factors under the control of the operator* such as:
- the range scale selected
- the correct adjustment of controls
- ambient light

(h) *Other factors* which influence positional accuracy are discussed in more detail later, here they may be summarized as:
- mechanical excellence of the equipment
- stability of the radar platform
- accuracy of inputs − such as from compass and log

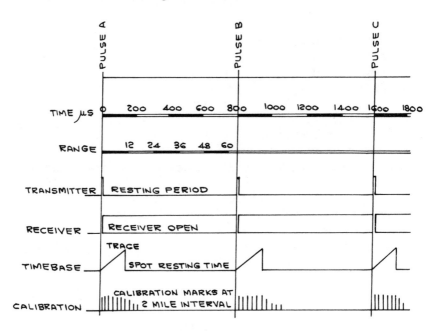

Figure 1.2 Time function diagram

Figure 1.2 shows a timing diagram. This relates such factors as PRF, range in use, pulse length and shows the interscan, or resting period of the timebase.

1.3 How many transmitted pulses strike the target?

A small calculation relating the scanner rotation speed, aperture size of the aerial and PRF of the transmitter demonstrates the number of pulses which fall on a target, subtending an angle of one degree, during every scanner revolution.

Aperture size of 2 metres with a pulse of frequency 10,000 MHz produces a beam approximately one degree wide in the horizontal plane.

If the scanner is rotated at 20 RPM, it completes one revolution in 3 seconds and hence takes $1/120$ seconds to pass through one degree. Since the target itself subtends one degree and the beam is one degree wide, it will thus be receiving energy from the radar during a period of $2/120$ seconds ($1/60$th).

6

Suppose the transmitter is designed to send the energy out in pulses at the rate of 1200 pulses per second, then the target will receive $^{1200}/_{60}$ pulses (20 pulses) during each revolution of the scanner.

THIS FIGURE SHOWS HOW AN ECHO IS BUILT UP BY THE RANDOM RETURNS ON SUCCESSIVE SWEEPS CAUSING BRIGHT "SPOTS" IN THE PHOSPHOR.

THE INITIAL STRIKE FADES SLOWLY WITH TIME, LATER RESPONSES FILL IN SPACES TO PRODUCE A "SOLID" ECHO.

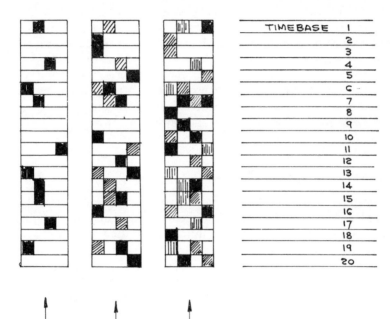

Figure 1.3 Paint build up

Of these (20) pulses striking the target, some may be 'lost'. On striking the target, energy is re-radiated, and some returns towards the original transmitter. Unfortunately the polarization of the re-radiated

7

energy may not be suitable for the receiving aerial to detect, or the direction in which the bulk of the re-radiated aerial energy is reflected may be away from that of the aerial. In either case the signal is too weak to rise above the receiver noise, and so cannot contribute towards the echo painted on the display. Statistically, a usable return from a poor response target may occur in approximately 3 out of 8 transmissions, and so only 7 or 8 detectable echoes will return from the 20 pulses which were sent out. They will occur randomly spaced, and with slight variations in range. The range variations may be due to infinitesimal steps in the trigger pulses, which may not start the spot each time from exactly the same point at the centre of the display; or it may be due to variation in echo strength changing echo shape, or, most likely, to a phenomenon known as glint. A target as complex, in radar response terms, as a ship will not reflect each pulse from the same point on every successive occasion, the apparent movement of the reflecting point is termed the glint of the target.

Figure 1.3 shows how an echo is built up on each successive transmission and how the successive scans correlate at three second intervals to form a 'paint' in the phosphor.

1.4 Unwanted echoes

The idea of the radar screen being comprised of some 3,000 invisible 'spokes', on which the ranges of targets are indicated by a brightened spot, is useful in considering the influence on the screen of unwanted echoes.

The brightness of each individual paint is obviously dependent on the strength of the response from the target. Strongly reflecting targets, even if only small, will cause a brighter spot than a poor response target. Successive paints in the same location on later timebases, or later sweeps of the antenna, will brighten the appearance of the echo so that useful targets appear. The returns from such unwanted targets as rain drops or sea waves closer to the ship, while possibly weak in themselves, occur on so many paints and in so many directions that they build up echoes over wide areas. Frequently these unwanted echoes may be strong enough to build up such a bright area on the screen that the wanted targets, because of lack of contrast and the effect of limiting or saturation, are lost to the observer.

Note: The way in which rain absorbs energy is another reason why targets are lost if they are in (or beyond) rain, simply because the radar does not receive enough energy to make a detectable response, but in this event there is little that the observer can do to improve the situation. The only safe behaviour in heavy rain is to assume that the effective range of the radar has been radically reduced and adjust the ship's speed accordingly.

1.5 Cleaning up unwanted echoes

Because of the wide range of targets which are likely to be picked up by a radar, and the need to amplify all signals enough to pick out the poor response of a distant buoy, the massive return from another ship when at short range may be amplified to a point where it could damage the screen. It is necessary to protect the tube phosphor from being defocussed. To do this a video limiting circuit is employed which cuts off the 'top' of signals which are too strong, while leaving other signals unaffected, see Figure 1.4. The penalty of using this circuit is apparent when strong unwanted signals, such as sea clutter, occur in the same area as a wanted signal from a ship. Figure 1.4 shows that, although the ship response is much stronger than the sea returns, the two signals are the same brightness on the screen and the ship is 'lost' in the clutter.

In this type of situation, where the wanted echo is as least as strong as the unwanted clutter which surrounds it, a variety of methods of cleaning up the clutter and displaying the wanted signals are possible. The methods fall into two main groups; the improvement is achieved by:

(a) cleaning up the raw signal before it is displayed on the screen
(b) extracting the whole signal train, manipulating it in a microprocessor and then synthesizing a picture from the data in the processor.

1.5.1 Analogue methods

The first method operates in the amplifier section of the receiver. By reducing the gain, that is the amount of amplification which all signals receive, the weaker signal can be reduced below the limiter setting and will show a weaker response. This technique is carried a step further by using an automatic gain reduction which is correlated to the range; this is the clutter control, which decreases the receiver gain appreciably when handling targets at the short range where most interference from sea response is likely to occur, and restores it gradually to ensure that targets at longer range obtain full amplification.

Another analogue approach is to use a differentiator circuit to extract targets from within clutter or rain. The technique of this circuit is that

9

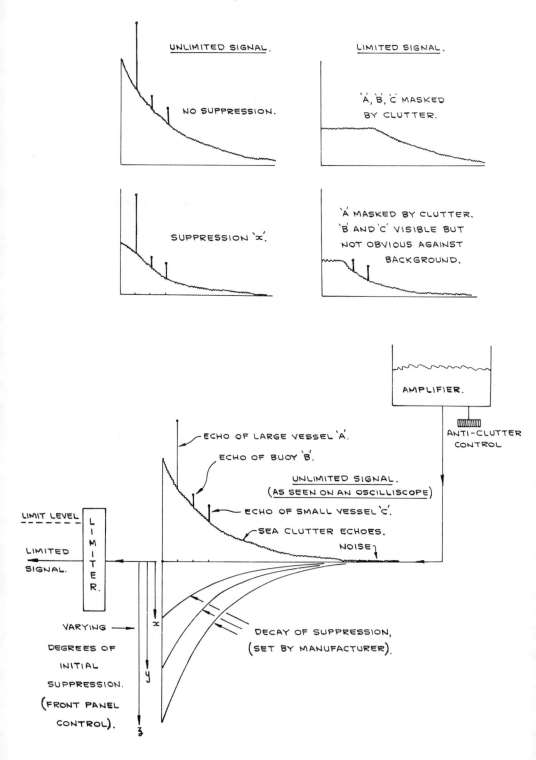

UNLIMITED SIGNAL.

NO SUPPRESSION.

LIMITED SIGNAL.

'A', 'B', 'C' MASKED
BY CLUTTER.

SUPPRESSION 'x'.

'A' MASKED BY CLUTTER.
'B' AND 'C' VISIBLE BUT
NOT OBVIOUS AGAINST
BACKGROUND.

AMPLIFIER.

ANTI-CLUTTER
CONTROL

ECHO OF LARGE VESSEL 'A'.

ECHO OF BUOY 'B'.

UNLIMITED SIGNAL.
(AS SEEN ON AN OSCILLISCOPE)

LIMIT LEVEL

LIMITER.

ECHO OF SMALL VESSEL 'C'.

SEA CLUTTER ECHOES.

NOISE.

LIMITED
SIGNAL.

VARYING
DEGREES OF
INITIAL
SUPPRESSION.
(FRONT PANEL
CONTROL).

DECAY OF SUPPRESSION,
(SET BY MANUFACTURER).

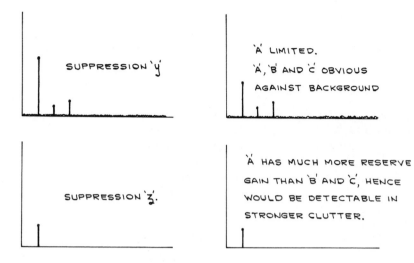

Figure 1.4   Control of sea clutter

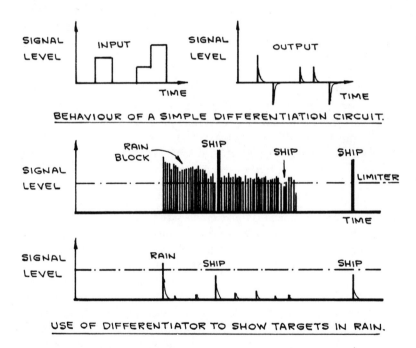

Figure 1.5   Use of differentiator (FTC) to detect target in rain

it operates to pass on to the display, not the train of signals as they appear, but a train of pulses which describe points at which changes in echo strength occur. The effect of this circuit is shown in Figure 1.5.

1.5.2 Digital method

The alternative approach is often adopted in addition to the analogue techniques so far described. In the second method more sophisticated circuitry is used to take the signal information from the amplifier, before it approaches the limiter or the tube, and to convert it into a block of binary data. A digital processor operates on this data to adjust it before it is replayed on the screen. Algorithms, or logical rules, are applied to the cleaning-up problem in place of the intelligent handling of the controls previously required of the man.

Actual techniques vary considerably from manufacturer to manufacturer; here, broad ideas are considered to demonstrate the type of solution adopted.

A switch register is a logic device, which effectively consists of a block of switches, side by side. Each switch can be changed from the 'off' into the 'on' condition on receipt of two simultaneous pulses. The arrival of a single pulse will not change the state of the switch. Suppose 32 such switches in a register are all connected to the signals returning on a timebase at the output of the amplifier; the other side of the switch is connected to an oscillator which 'clocks' each switch, one after the other. If a clock pulse and a signal occur at the same time the switch is left in the 'on' position; if there is no signal when the clock pulse occurs the switch will be left off. Hence the absence or presence of echoes on the timebase is mirrored by the 'off' and 'on' status of the switches. It is usual to denote the 'on' condition by a one and the 'off' condition by a zero; the timebase is thus represented to a computer processor as a binary number.

Figure 1.6 shows a typical timebase and switch register, a calculation is needed to show the significance of the various quantities involved. As shown, it is intended to copy just over three miles of timebase into the 32 cells, or bins, of the register. Reference back to the time diagram (Figure 1.2) shows that energy will have echoed back from 3.2 miles in, $3.2 \times 12.3 = 40.0$ microseconds. Thus the 32 switches must be 'clocked' at equal time intervals in the 40 microsecond period, or, every clock pulse must occur at $^{40}/_{32} = 1.25$ microseconds after the previous one. To achieve this rate the oscillator providing the clock pulses must run at 800,000 cycles per second (0.8 MHz). At the same time, since 32 bins accommodate 3.2 miles, every bin represents 1 cable of the timebase; this is the effective accuracy of this arrangement.

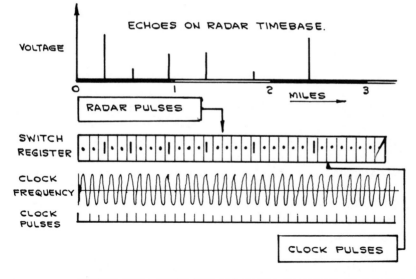

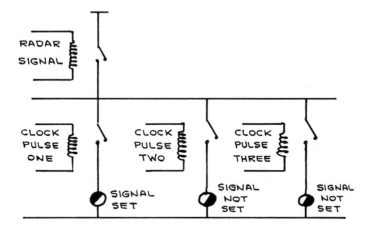

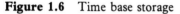

Figure 1.6   Time base storage

To increase the accuracy, say to ¼ cable bins, it is apparent that the number of bins must be increased to 4 × 32 = 128 and the clocking oscillator be increased in frequency to 4 × 0.8 MHz (i.e. 3.2 MHz). To

13

increase the range of the timebase which can be stored it is only necessary to increase the number of cells in the switch register while maintaining the frequency of the oscillator. Thus, to store 25 miles of data at ¼ cable (150 ft) accuracy requires approximately 1,020 cells, clocked at 3.2 MHz. In practice, clocking at higher frequencies introduces difficulties and this limits the range accuracy of data which may be stored, but techniques which use a number of registers in parallel and then phase the connection of the timebase to each line allow a high accuracy with a comparatively lower clocking frequency.

Clutter control can be achieved by setting a threshold value. Only signals stronger than this threshold value will be accepted into the switch register. In simple systems the threshold may be a fixed level, but this inflexibility means that either some poor response signals are lost, or some clutter has to be tolerated. More sophisticated systems use a dynamic threshold level, which is continuously adjusted by the strength of the clutter being obtained to ensure that the threshold allows all signals, but very little clutter, to break through.

1.5.3 Regenerating a synthetic picture

After the data is stored a synthetic radar picture can be generated from the processor. Such a picture has many advantages over the raw radar. Typical facilities are:

(a) *Timebase to timebase comparison* noise and unwanted echo, (clutter and interference) rejection
(b) *Echo shaping* (more obvious echoes from poor responses)
(c) *Brightness levelling* (all echoes at maximum brightness)
(d) *Constant speed timebase* (bright/daylight displays)

If one or more timebases are stored in the processor registers it is comparatively simple to compare the signals in each, and to reject by a logical decision any of the signals which are random noise or interference, Figure 1.7. Any rain or sea clutter returns may also be rejected.

On the other hand, where it is evident that a target exists, but a response has been missed on one timebase, a cell may be filled to block out the echo.

Again, since an echo, irrespective of its actual strength, is represented by one digit it is possible to build all echoes on the display at the same level of brightness. Other systems which set more than one threshold at different levels, to store a stepped value of echo strength, are able to regenerate the picture with several levels of displayed brightness.

Finally, it is possible for the speed of the spot crossing the cathode ray

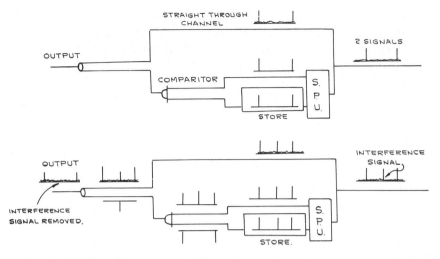

Figure 1.7 Interference suppression

tube to be the same, whatever the range scale adopted. It is only necessary to clock the data out of the register at a rate which depends on the range which the observer wishes to display.

To return to the previous example, the information stored at ¼ cable accuracy in 1,024 cells in fact covers 24 miles. The timebase for the spot deflection on the screen can be constant at a little less than 1/PRF, say, 1000 microseconds. To show a 12-mile picture it would only be necessary to clock out the first 512 cells in 1000 microseconds, or to use a clock oscillator at 500 kHz. For a 6-mile picture clock out the first 256 cells in the same 1,000 microseconds, or 250 kHz. This means that although responses are received and stored in 'real time', they are read out of store and displayed at a different rate; the shorter the range scale, the slower the 'read out' rate.

The faster clocking capabilities of more modern circuits allows a slightly different, but more accurate approach. Using this technique the time taken for the spot to cross the tube remains the same on all ranges, while the clocking out rate also remains constant. To accommodate the different ranges, the clocking in pulses are adjusted to fill the whole register with the range in use. Approaching the problem this way permits an increasing range accuracy resolution as range scale decreases. For example, in the previous case where 1024 bins are used and the spot deflection must be completed in 1000 microseconds, the constant clocking out rate must be approximately 1 MHz (one cell per microsecond). Used on the 24 mile range scale, time for energy to return from 24 miles is approximately 300 microseconds so clocking in

15

rate will have to be at 3.3 MHz and cell accuracy will be about .24 cables. When used on the 3 mile range scale, the clocking rate would have to be eight times as fast, but the cell accuracy would also be eight times higher.

Chapter 2

Extraction and Analysis of Data

2.1 How is collision avoidance information obtained from echo movement?

The radar is only capable of displaying the immediate range and bearing of an echo. Any further intelligence which is required to make collision avoidance decisions must be inferred from the movement of the echo; the ratio of change of its range and bearing. This is invariably achieved by plotting the history of the echo movement, either manually on paper or reflection plotter surface, or by using one of the several proprietary appraisal aids which are available.

Two levels of information are required:

(a) *the risk of collision* as indicated by the relative track to determine need to act, and

(b) *the true motion* to aid choice of action

Either of these two modes may be displayed on the radar and the motion tracked, by simply recording the positions of the echo at discreet time intervals. To obtain the additional information requires some level of vector plotting to analyse the motion tracked.

2.1.1 Relative information

Figure 2.1 describes the conventional form of data extraction. The relative motion has been tracked and the quantities of target real motion have to be found. Several plots taken at equal intervals of time are recorded and the projection of a line through these, past the centre of the display, then permits measurement, or estimation, of the miss distance. (Nearest approach). The point where the apparent track touches the miss circle is termed the closest point of approach (CPA).

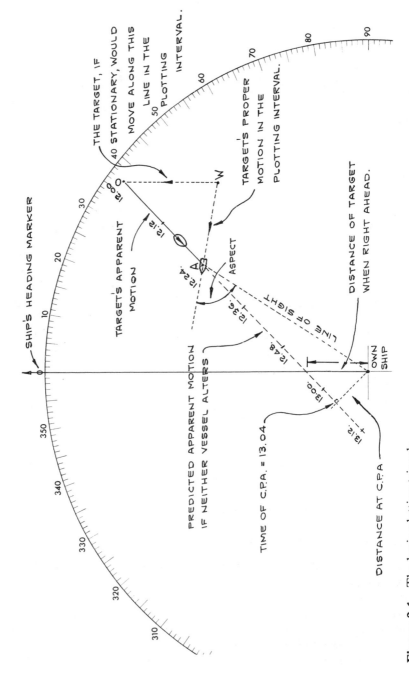

Figure 2.1 The basic plotting triangle

Indication of the urgency of the situation can be obtained by extrapolating the apparent speed of the echo into the distance to go to CPA (i.e. Time to Closest Point of Approach or TCPA).

2.1.2 True information

True motion information is found by applying the motion of own ship, during the plotting interval, to the relative track. The vector resolved shows the course of the target relative to own course and the proportional speed. Exact target speed can be found by simple calculation, but speed ratio is often sufficient for manoeuvre purposes, and this may be obtained by comparison of own ship vector length and target vector length.

Aspect, the angle between the target course and the line of sight to own ship is usually the most significant piece of information available.

2.1.3 Target manoeuvres

Figure 2.2 shows the changes in relative track which occur when the target manoeuvres. Either speed change or course change will cause a change in relative motion; the only sure way to identify the actual action by the target is by plotting before and after the change is established. It is important to note the unstable nature of the track during the time that the manoeuvre is taking place. The period of uncertainty will depend on:

(a) *the original apparent speed*
(b) *the degree of alteration* of course and/or speed chosen
(c) *the rate* at which the alteration is made
(d) *the plotting interval* being used

2.1.4 Own ship manoeuvres

Figure 2.3 describes the construction needed, when own ship makes a manoeuvre, to determine what track the echo will follow. This technique is termed 'the trial manoeuvre' since it is normally carried out before the actual action is implemented. Again, in practice, during the period that a manoeuvre by own ship is being executed, the relative track of the target is difficult to decipher. Quick and positive action will obviously minimize this problem.

Exercises and more detailed instruction in plotting exercises appear in Appendix III.

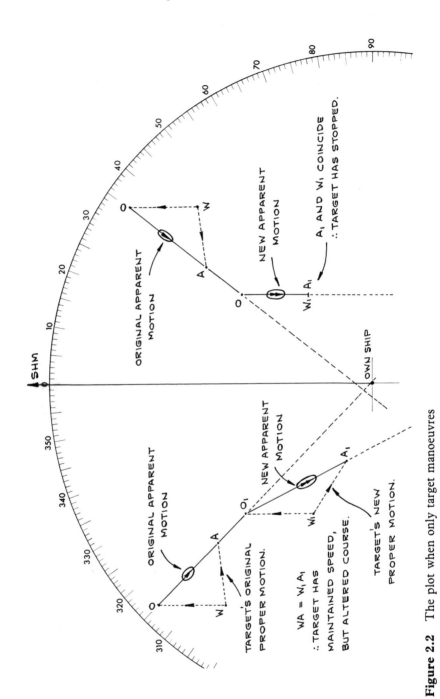

Figure 2.2 The plot when only target manoeuvres

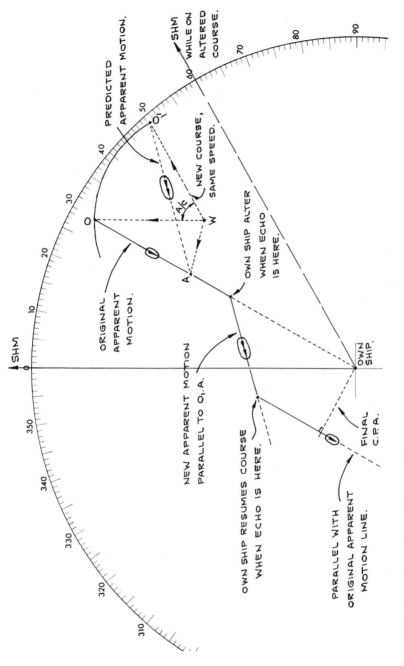

Figure 2.3 The plot when only own ship manoevres

2.2 What is the anti-collision loop?

Figure 2.4 describes a flow diagram which outlines the 'anti-collision loop'. Targets and own ship combine geometrically to form 'an encounter' or, if more than two ships are involved, 'a situation'. The observer uses sensors, eyes in clear weather or radar in fog, to examine the immediate scene for measurable data. This data is measured, by range, bearing and time for example, then stored in some data base which is conventionally a 'plot' or track on a plotting surface; in less demanding conditions the range and bearing simply may be noted. The stored data is analysed to describe the relative motion, the CPA and

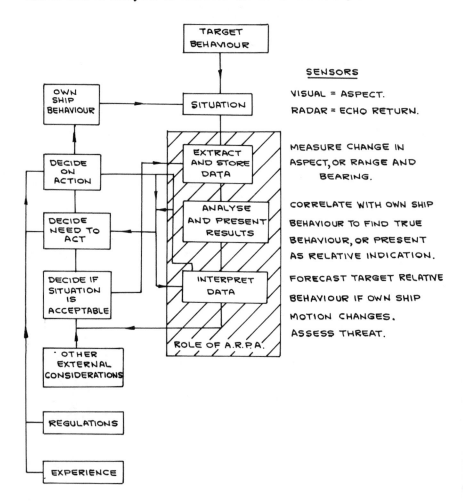

Figure 2.4   The anti-collision loop

TCPA. The elements of own course and speed are then introduced to establish the proper or true motion of the target. This information is displayed in some form, such as an arrow on the plot, and the geometry of the situation is interpreted. The loop now enters the decision-making phase when the observer combines the displayed information with his experience, the applicable Regulations, the condition of his ship, the geographical position, etc., to decide if any action is needed. If a manoeuvre is necessary the observer may 're-enter' the interpretive block to attempt a trial manoeuvre before taking action.

Finally, own ship alters, which changes the situation, and the loop is re-circled to measure the success of the action.

2.3   What is the role of ARPA?

The role of the ARPA is shown in the shaded part of Figure 2.4. As presently envisaged, all that the system will do is to take data from the radar sensor, store the data in a database, operate on the database to obtain the relevant information and then display the data in a digested form (such as vectors) to the observer. The system will usually allow input of trial conditions to show the influence of trial manoeuvres.

The role of the ARPA is thus seen to be the extraction and analysis of data, relieving the man of a tedious, time-consuming task. The ARPA does not usurp the decision-making role of the mariner, although the increased level of information and rapid appreciation of trial manoeuvres make it a powerful tool in assisting the man to make the decision.

2.4   How does the radar data get into the processor?

Radar returns in the amplifier are transferred into switch registers in a similar way to that described for cleaning up the picture, (see page 9).

Generally, one of two approaches are considered appropriate. Either *automatic acquisition* is used to collect all data which would appear on the radar screen so that targets are selected and tracked without the intervention of the mariner, or *manual acquisition* is used, which requires the mariner to define the selected targets he wishes to track. Once tracking is initiated, whatever the system, the tracker continues to follow the target until the track is deleted by the mariner, or the target is 'lost'.

While *automatic acquisition* will reduce some of the work load, it does mean that there are usually more vectors displayed than are really necessary, and in heavy traffic there can be doubts about the priority parameters.

Manual acquisition suffers from the time taken to 'acquire' and 'delete' targets, but it does mean that acquired targets are those which the navigator really wants. While manual acquisition is a must in the IMCO specification, automatic acquisition is permissible and is available from most manufacturers and fitted as standard by many.

2.4.1 Typical tracker design

Figure 2.5 describes a typical automatic tracker. A number of switch registers are available to store the data. 256 cells will accommodate over 25 miles at 1 cable accuracy. The prime register is clocked in during each timebase. During the rest (interscan) period the prime register is cleared by being transferred into register one. At the same time register one is shifted into register two, and two into three, etc. The data in the last register is wasted. During the transfer a comparison of the content of the same cell in all registers is made; if there are more than m out of n cells with a 'one' in them, a cell in the same position in a 'hit register' is set to 'one'. All the cells along the register are compared in turn. This filter helps reduce the number of false echoes likely to be encountered, and at the same time allows the probability of return from weaker targets to be taken into account.

2.4.2 Shaft encoding

To complete the data store it is necessary to record the bearing, along with the relevant range data stored in the 'hit register'. This is achieved by building a shaft encoder on to the scanner rotation mechanism. A simple shaft encoder having 5 concentric sets of segments is drawn in Figure 2.6; this can resolve bearing accuracy down to $11\frac{1}{2}$ degrees, or $\frac{1}{32}$ part of a circle. Practical encoders can be built with 9 or 10 segmented rings which will give a bearing accuracy down to $\frac{1}{512}$ or $\frac{1}{1024}$ part of a circle respectively, while 12 digit encoders giving an accuracy of $\frac{1}{4096}$ part of a circle are common. The code output by the shaft encoder can be associated with the data in the 'hit register' to build up a matrix of ranges and bearings of the target locations.

In a manual tracker, a gate can be positioned over the wanted target. This has the effect of starting the 'clock' on the register just before the target is reached and stopping it shortly after, so that only a few cells of the switch register are loaded. Similarly only a few selected codes from the shaft encoder are used. The hit matrix will now only hold the range and bearing data of the wanted targets.

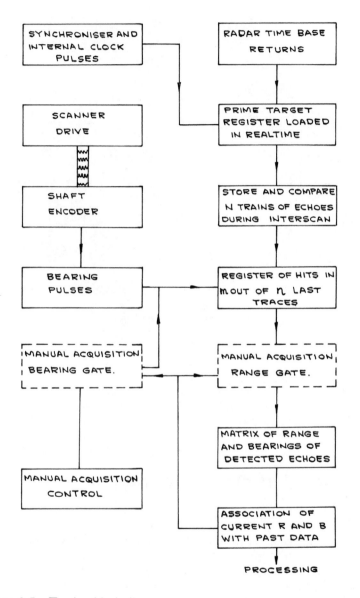

Figure 2.5 Tracker block diagram

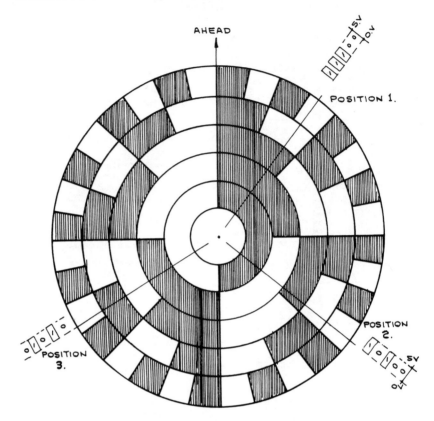

Figure 2.6 Mechanical shaft encoder

Shaded areas are connected to + 5 volts. Clear sectors are connected to 0 volts. The rotating wiper carries five brushes which have signals, as shown at position 1, 2, 3. These may be decoded as 1 and 0 to form a code. The code changes continuously as the wiper rotates in synchronism with the scanner.

2.5   How does the ARPA deal with the basic data?

Auto acquisition is technically difficult to achieve, and also poses greater problems in processing. Only a limited number of targets can be dealt with by the type of computer usually available in a commercial tracker, so there must be some degree of polling or prioritizing to determine which targets can be dropped or rejected. Targets subtending angles greater than two degrees (at longer ranges), for example, are almost certainly too big to be ships, but this is only a small part of the problem. In a wholly automatic system there is a constant need to perform an association process, to connect echoes detected on one scan with those found on the previous scan, and only this allows old targets to be followed and new contacts to be evaluated.

Once the echoes are associated and established in the tracking file the processor must operate on that data to produce the apparent motion track. Calculations in polar terms, that is range and bearing, are difficult:

(a) Range and Bearing rates are not constant for straight tracks
(b) The spatial resolution varies with range (i.e. it is geometrical)

and it is usual to convert the data into cartesian values of eastings and northings. What the processor must now do is establish the rate of change of northings and eastings of the target. The method adopted by the human operator, that is, drawing the best fit line through the sufficient positions plotted has disadvantages in a computer solution, although least squares fitting methods have been used.

2.5.1 Analysis of data

One disadvantage of the least squares method lies in deciding on how many basic positions to include in the analysis. If there are too few, the line develops a wander − if there are too many, the line becomes rigid and real changes in apparent motion are difficult to identify.

A better approach is obtained by using a type of regression analysis. Figure 2.7 shows how successive pieces of information can be used to improve the forecast of the next position in which the echo is expected to appear. The two quantities being calculated over a period of 10 or 15 scans can be labelled α and β. The size of β is really a measure of confidence in the tracking and the smaller this value becomes, the more precise the answer will be. In this way, it is possible to establish a feedback loop in the processor progressively reducing the size of the tracking gate used in the manual system. The advantages of a reduced tracking gate are:

(a) *early identifications of changes* in apparent motion
(b) *less likelihood of 'target swop'*
(c) *good rate aiding*, i.e. the ability to track a target through clutter, rain, or when the signal is otherwise intermittent

Once the relative motion, as a rate of change in Easting and Northing has been established, application of the elements of own course and speed, suitably resolved into North and East components, can be applied in the processor to generate the true course and speed of the target.

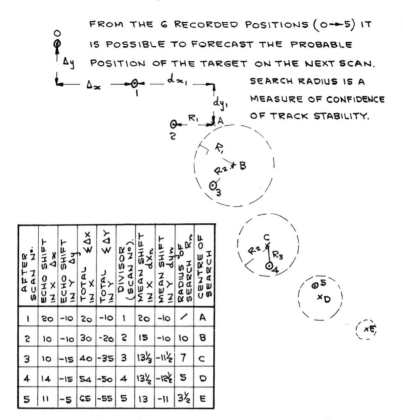

FROM THE 6 RECORDED POSITIONS (0→5) IT IS POSSIBLE TO FORECAST THE PROBABLE POSITION OF THE TARGET ON THE NEXT SCAN. SEARCH RADIUS IS A MEASURE OF CONFIDENCE OF TRACK STABILITY.

AFTER SCAN N°.	ECHO SHIFT IN X Δx	ECHO SHIFT IN Y Δy	TOTAL IN X WΔx	TOTAL IN Y WΔy	DIVISOR (SCAN N°)	MEAN SHIFT IN X dXn	MEAN SHIFT IN Y dyn	RADIUS OF SEARCH Rn	CENTRE OF SEARCH
1	20	-10	20	-10	1	20	-10	/	A
2	10	-10	30	-20	2	15	-10	10	B
3	10	-15	40	-35	3	13⅓	-11½	7	C
4	14	-15	54	-50	4	13½	-12½	5	D
5	11	-5	65	-55	5	13	-11	3½	E

Figure 2.7 Rate aiding
Note The running mean is stored to project the probable centre of the next search circle. The divisor in the sixth column can be limited to a number of scans or controlled by some value of search radius. The radius of the circle can be increased if echo is not found.

Similarly when a trial manoeuvre facility is required, the processor can be fed with the intended components of own ship's motion and the forecast of target motion under these circumstances can be produced.

2.6 How does the ARPA display information?

The ARPA can present the data from the processor in a variety of different ways. The most obvious, and perhaps simplest, is a display of information printed on a VDU (video display unit). Alternatively, the data may be converted into graphics and can be displayed as vectors in either true or relative mode. More sophisticated displays may generate more information like the probable area of danger or collision point. Future equipments may show more and different types of data.

The mode of display when vector graphics are selected may be on a separate graphic terminal which also incorporates a synthesized picture of the terrain. Others may use the interscan period of the raw radar to write the graphical data on the 'live' radar picture. On this type of display, the picture may or may not have been artificially cleaned. The choice of method seems to be equally divided between different manufacturers, there are advantages and disadvantages to both systems: –

(a) *the wholly synthesized display* may have lost echoes because of the cleaning process
(b) *the raw radar with 'data overwrite'* may carry more data than is strictly necessary to the observer but will not suffer from weak target loss as a result of processing

2.7   How is the ARPA tracker used in practice?

Consider a target which is manually acquired by means of the joystick, and placed within the tracking gate. The gate automatically centres itself on the target and the computer tracks the gate. As the scanner comes around, the gate will reposition itself if the target has moved, but since each step is finite, and fairly coarse, the target will 'appear' to move in jumps – see Figure 2.8.

2.7.1   Target loss, rate aiding and target swop

So far it has been assumed that on each scan, a target response is received, detected and stored, but if on a particular scan no response is received, then the gate does not know where to go and so remains where it is.

If there is only a temporary loss, and if when the response is again received, the target is still within the gate, tracking will be resumed. If on the other hand, when the response is again received, the target is outside the gate, the target will be 'lost'.

This problem can be overcome by 'rate aiding'. Once tracking has commenced and the target's movement has been established, then at each scan, the gate is moved to where the tracker expects the target to be. The ARPA is required to be able to continue to track an acquired target when as many as 5 out of 10 consecutive responses have been lost.

29

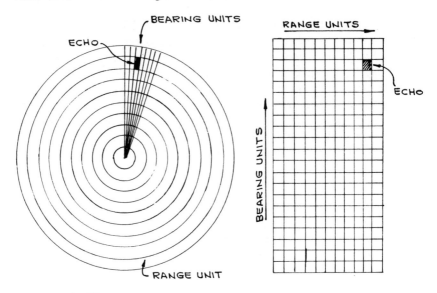

Figure 2.8 Picture storage
This figure shows how the position of a target in range and bearing can be placed in a matrix defined by range and bearing units. Note how the bearing precision improves as range decreases.

Consider the situation as depicted in Figure 2.9, where two targets – one tracked, the other not – are likely to come within close proximity of each other, and at about that time, the responses from the tracked target are intermittent (e.g. in shadow of the other vessel).

If the 'tracked' target alters course, and the effect of rate aiding now moves the gate on in such a way that it now encircles the other target, it will continue to track that target, and target swop has occurred.

Target swop is also possible when 2 targets fall within the same gate and is therefore more likely to occur in high density traffic and in port approaches. On at least one ARPA, it is possible to deliberately reduce the gate size (Harbour Mode) in an attempt to overcome the problem, while another automatically adjusts each gate *individually* to match the size of the echo it is tracking.

2.7.2 Prioritization

In any system, whether it is manual acquisition or, more probably, automatic acquisition, saturation of the core space – the working area of the processor – will mean that the computer will have to decide which targets must be tracked and which must be rejected. The rules which govern rejection of targets are based on measurable parameters such as:

(a) *the bearing*
(b) *the range*
(c) *the miss distance*
(d) *the time* of the miss distance

Quite complicated programmes can be written which weight these variables to give the best answer. Sometimes the operator of the equipment is allowed to change the weighting of the various components but in many cases the rule is transparent to him, that is he

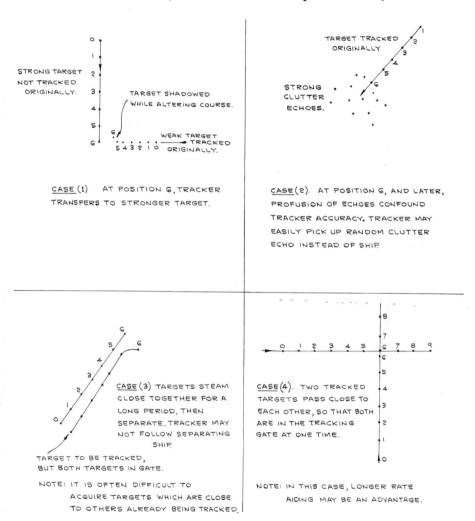

CASE (1) AT POSITION 6, TRACKER TRANSFERS TO STRONGER TARGET.

CASE (2). AT POSITION 6, AND LATER, PROFUSION OF ECHOES CONFOUND TRACKER ACCURACY. TRACKER MAY EASILY PICK UP RANDOM CLUTTER ECHO INSTEAD OF SHIP.

CASE (3) TARGETS STEAM CLOSE TOGETHER FOR A LONG PERIOD, THEN SEPARATE. TRACKER MAY NOT FOLLOW SEPARATING SHIP. TARGET TO BE TRACKED, BUT BOTH TARGETS IN GATE.

NOTE: IT IS OFTEN DIFFICULT TO ACQUIRE TARGETS WHICH ARE CLOSE TO OTHERS ALREADY BEING TRACKED.

CASE (4). TWO TRACKED TARGETS PASS CLOSE TO EACH OTHER, SO THAT BOTH ARE IN THE TRACKING GATE AT ONE TIME.

NOTE: IN THIS CASE, LONGER RATE AIDING MAY BE AN ADVANTAGE.

Figure 2.9 Target swop

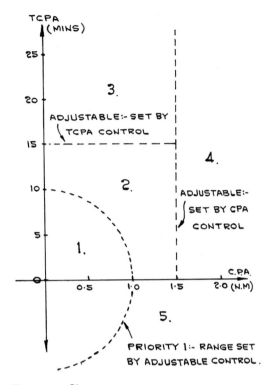

Figure 2.10   Threat profile

cannot see how it is working. Figure 2.10 shows a typical selection algorithm which might be used in an automatic acquisition system.

In the ARPA Specification it is required that the priority algorithm is made known to the user. In most cases it is merely 'The nearest "n" targets'.

Chapter 3

ARPA Display Facilities

3.1   General introduction to ARPA types

3.1.1   Early days

The earlier users of radar (1945 – 7) were at a considerable advantage during periods of reduced visibility because of their superior knowledge of the presence of other ships. With only a few ships fitted with the equipment and the remainder resorting to the traditional slow speed in fog, there was no difficulty for the fitted ships to use their knowledge of the other ships' position to pick a course clear of danger. As fitting became more widespread, this advantage was eroded, but the bad practices learned during the early years were slow to disappear and the number of accidents involving radar equipped ships mounted steeply. The interpretation of the relative motion described on the PPI of the radar proved difficult to mariners who had spent all their watchkeeping experience dealing with the 'aspect' of the target in the visual sense. The problem of translating the relative motion supplied by the radar to true motion became the cornerstone of radar development, first by introducing courses which taught the mariner how to 'plot' on paper sheets, then by the appearance of the anti-parallax reflection plotter to allow plotting on the tube face directly. Other electro-mechanical transfer methods also appeared, usually involving the rotation of a gantry in sympathy with the bearing cursor, and the movement of a trolley on the gantry controlled by the range marker. Final marking was done by a pen or stylus on special paper.

3.1.2   True motion displays

Later a technological solution was sought by the introduction of true motion. In this the motion of own ship was mechanically resolved and used to displace the centre of the radar picture across the screen at some scaled speed, and so show the true tracks of the target. However, true motion still required some level of plotting to be carried out.

3.1.3   First off-screen recording

The first attempt actually to record the history of the echo motion was made in the 'Photoplot' radar which photographed the track over a period of time and then projected the historical track on a flat plotting surface. Other approaches, which usually used the idea of placing fixed markers on the echoes, followed. Markers which indicated the collision track, and assumed that divergence by the echo from the track equalled safety, were a big practical step forward because they allowed true motion to be watched but relative motion to be assessed. This philosophy of having both data simultaneously available is still apparent in many ARPAs.

3.1.4   The Predictor

Up to this time no attempt had been made to extract the echo itself from the display. The first equipment to do this was the Predictor which formed the first step in the ARPA direction. In this equipment the whole picture was taken off the radar and stored on video tape. A small processor running in the equipment also stored the motion of own ship, and allowed pictures of up to six minutes history to be replayed in either relative or true motion. Figure 3.1 shows a typical Predictor picture of the two modes. This equipment was unique in a second way since it allowed a forecast of the effect of a trial manoeuvre to be described, as in Figure 3.2.

3.1.5   ARPA facilities

Finally, the first true ARPA appeared, a system able to extract the signals from the radar head and pass them to a digital processor. Once the data is within the processor these equipments can use a variety of facilities to present information to the observer. These facilities include:

1 Relative vectors
2 True vectors
3 Points of collision
4 Predicted areas of danger

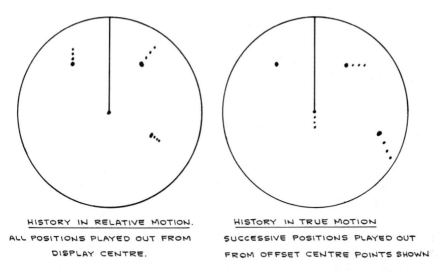

Figure 3.1   Presentation of history

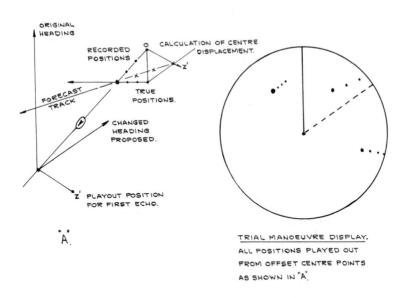

Figure 3.2   Trial manoeuvres facility

5 History tracks
6 Trial manoeuvres
7 Digital Data output
8 Navigation lines and limits
9 Operational warnings
10 Equipment warnings
11 Rejection boundaries

3.2   How is an ARPA used?

Here it is perhaps convenient to consider one example of an ARPA display as depicted in Figure 3.3. This is a synthesized display having automatic as well as manual acquisition, the display being an addition to the normal radar display. The ARPA is connected to the radar, from which it automatically extracts data, processes it, and displays it along with graphics and possibly alpha-numerics. A computer forms the heart of the system which plots the targets and displays the vector associated with each tracked target.

Having first set up the ARPA display (as for normal radar display), select:

(a) *Range scale* – e.g. 12 miles
(b) *Plot* – Relative (or True)
(c) *Mode* – North up (Head up or course up)
(d) *Mark the targets* to be tracked (using joystick and gate)
(e) *Set the 'Vector length'* – in minutes
(f) *Check the course and speed input*

3.3   How is basic information displayed as a vector?

After some 10 sweeps of the scanner, the computer has enough data (10 plots) from which to determine the target's motion. Attached to each tracked target will appear a vector, whose length will be indicative of the mode speed. The end of the line indicates where the target will be in 'x' minutes time as set by 'vector length'.

In the relative vector mode, the 'vector length' control can be used to extend the vector lines to pass the display origin, and thereby indicate CPA and TCPA. The direction of each vector will indicate the apparent or relative motion of the target, depending on the chosen mode. A typical relative situation is depicted in Figure 3.4(a).

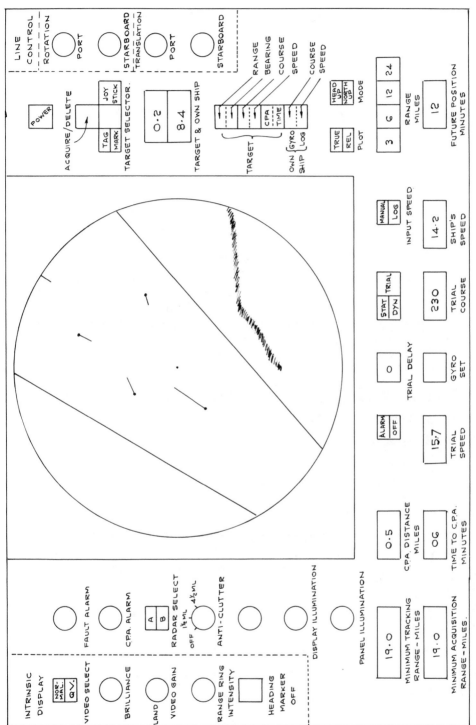

Figure 3.3 A typical ARPA

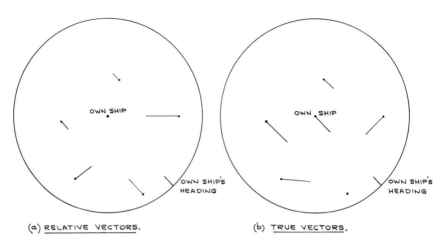

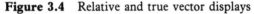

Figure 3.4 Relative and true vector displays

With a knowledge of 'own ship's' course and speed, the computer can determine the true vectors of the targets as is shown in Figure 3.4(b). Note, that 'own ship' also has its true vector displayed, with a vestigial heading marker at the tube edge. As in the relative presentation, the vector lengths can be increased to indicate where all tracked targets will be at some particular time in the future.

3.4 How is numerical data relating to a particular target found?

By using the joystick and placing the gate marker ring over a particular target, data in numerical form relating to that target can be obtained:

(a) range and bearing
(b) course and speed
(c) CPA and TCPA

This data may be made to appear sequentially or simultaneously on a special data display. Alternatively, alpha-numerics may be used to make the data appear on the ARPA display, alongside the particular target.

3.5 What extra facilities are available?

3.5.1 Trial manoeuvre

It should be possible to simulate the effect of a manoeuvre – by 'own ship' – on all tracked targets.

This is done by feeding in:

(a) *the proposed course*
(b) *the proposed speed*
(c) *a delay* (if any) e.g. a/c at 0527.

The display can be made to indicate the effect of such a manoeuvre. The method of display may be either static or dynamic, in which case, the tracked targets and own ship are made to move at some 30 times normal speed. 'Own ship' will of course move at the 'proposed' speed in the 'proposed' direction, with 'own ship's' handling characteristics being taken into account.

Note: While trial manoeuvres are being simulated on the display, the computer continues its normal task of tracking all marked targets in the background.

3.5.2 Operational warnings

(a) *CPA warnings* It is possible to set limits of CPA and TCPA which, if violated by a tracked target, whether its vector actually reaches the warning area or not, will activate an alarm. The offending target will be indicated e.g. by a brighter than normal or flashing vector, or a special symbol.
(b) *Guard rings and zones* It should also be possible to warn the observer if any distinguishable target closes to a range or transits a zone chosen by the observer. The target activating the alarm should be clearly indicated e.g. a flashing marker. The guard ring(s) or zone(s) may be pre-set or adjustable. It is important to remember that targets already within the guard ring or nearer than a zone when they first appear will not activate the alarm. The existence of guard rings should not be regarded as an alternative to keeping a proper lookout.
(c) *Target lost* The ARPA should clearly indicate if a target is lost (see also rate aiding) with the last tracked position being clearly indicated.

3.5.3 Area Rejection Boundaries (ARBs, AEBs)

It is possible to place electronic lines on the screen which eliminate automatic plotting in selected areas. The lines are adjusted by 'rotation' and 'translation' controls. These reduce the load on the tracker when in proximity to a coast echo, for example.

Alternative systems provide automatic acquisition in zones which may be designated by range and sector controls.

Note: Manual acquisition avoids the need for ARBs.

3.5.4 History display

Data relating to all tracked targets are stored for some 8 minutes and 4 equally time-spaced past positions can be observed on request. This enables the observer to check whether a particular target has manoeuvred in the recent past, e.g. while the observer was temporarily away from the display on other bridge duties.

3.5.5 Equipment faults

Connections with other equipment
The connection of the ARPA to any other equipment should not downgrade the performance of that equipment. The failure of an input from other equipment, such as log or compass, should activate an alarm.
Performance tests and warnings
Self diagnosis should activate a warning in the event of ARPA malfunction. Also, means shall be available to check the correct interpretation of data against a known solution.

3.6 What alternative facilities are available on ARPAs?

3.6.1 Automatic acquisition

It is permissible for targets to be automatically (as well as manually) acquired, but where automatic acquisition is provided, the operator must be able to select the areas in which it operates.

Where automatic acquisition is provided, the ARPA should be able automatically to track, process and simultaneously display and continuously update the information on at least 20 targets whereas only 10 targets need be tracked if manual acquisition alone is provided.

Problems which arise with automatic acquisition:

(a) How does the ARPA distinguish 'ship' targets from, say, land, clutter and false echoes?
(b) If there are more 'ship' targets than the nominal 20 tracking channels, which 20 targets should be tracked and vectors displayed?

The first thing which has to be done is to 'clean up' the picture. This elminates 'Interference' caused by other ships' radars, reduces clutter (from sea or precipitation) and enhances the remaining echoes which may then be interrogated for size. Echoes subtending an angle larger than some pre-set limit are rejected. The operator may assist considerably by the use of such controls as the ARBs and 'Minimum tracking and Acquisition Ranges'.

Targets that now remain are potentially 'ship' targets and require tracking. ARPAs may maintain constant surveillance on some 200 ship-sized echoes and must track and fully plot a minimum of 20.

A problem arises in deciding which targets should be tracked and displayed; of some 200 targets being 'observed', which are the 20 most important from an operational viewpoint? It might be the twenty targets with least range. Alternatively it may be the twenty with the closest CPAs, or it may be thought important to take into account the closing rate. Maybe some combination of all three is considered essential. In any event the criteria for the selection of targets for tracking must be provided for the user. A typical priority diagram is given in Figure 2.10.

3.6.2 Tracking and acquisition limits

There may well be times when targets are close to 'own ship' (e.g. in traffic lanes and narrow channels) but present no real threat, and whose vectors may well clutter up the centre of the display. It may be possible therefore to set limits on the ranges at which targets are acquired and to which they are tracked.

3.6.3 Navigation lines

These may be set to assist with parallel indexing techniques. Some 10 or so pairs of electronic lines can be made available to set navigation limits, delineate danger areas or channel edges. These are used as parallel index lines in the normal way. Other systems provide facilities for entering and displaying more comprehensive data such as: way marks, traffic separation zones, or planned passage lines (see paragraph 10.4).

3.6.4 Ground lock

It may be possible to use one manual tracking channel to track a known 'stationary' target and use any detected movement as additional inputs to those from the course and speed sensors and in this manner, truly 'ground stabilize' the display. (It is important here to remember the dangers of misinterpretation which are likely when using a Ground Stabilized Display for collision avoidance, and therefore the inadvisability of using this mode of display under such conditions).

3.6.5 PPCs (Potential Collision Points)

From the basic plot of a target, it is possible to determine the course to steer (if speed is maintained) in order that a collision (or interception) will take place. It is possible to have these PPCs appear on the display and in this way, allow the navigator to avoid them.

Note: In a system where a target is not connected to its PPC, extra care must be taken to ensure that PPCs are related to the appropriate targets.

3.6.6 PAD (Predicted Areas of Danger)

It is a logical step from PPCs to indicate areas around these points into which a vessel should not go in order to ensure that some specified clearing range is maintained. These predicted areas of danger are a feature of the Sperry Collision Avoidance System and in earlier models appear as ellipses, in later models, as hexagons. These are considered in more detail in a later section.

Note: A line which is not a time-conscious vector joins the target to the centre of its ellipse(s). The centre of the ellipse is not necessarily the PPC.

3.6.7 Methods of testing an ARPA for malfunction

These usually take the form of self diagnositc routines (either automatically or manually invoked) with some indication of the unit or Printed Circuit Board which is found to be faulty.

Chapter 4

Collision Points and PADs

4.1 What is the concept of collision points?

When two ships are in the same area of sea, it is always possible for them to collide. The point(s) at which collision can occur may be defined and depends on:

(a) *The speed ratio* of the two ships
(b) *The position* of the two ships

In any pair of ships there is usually one which is faster than the other; the possibility that one is exactly the same speed as the other and will maintain that ratio for any period of time is remote enough to be disregarded for the moment.

The ship which is the faster of the pair will always have one and only one, collision point, since it can pursue the target if necessary. This collision point is always on the track of the target as shown in Figure 4.1.

The ship which is the slower of the pair may have two collision points, both of which must be on the target track. One exists where the slow ship heads towards the target and intercepts it, another exists where the slow ship heads away from the target but is struck by it. The two cases appear in Figure 4.2. Alternatively there may be no way for the slower ship to collide with the faster (even though the faster may collide with the slower) because it is just not fast enough to reach the target track, Figure 4.3.

Note: a critical in-between case of one collision point exists where the slow ship can *just* reach the track of the fast ship.

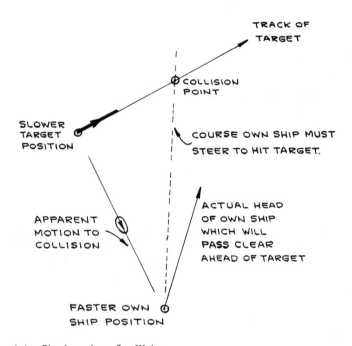

Figure 4.1   Single point of collision

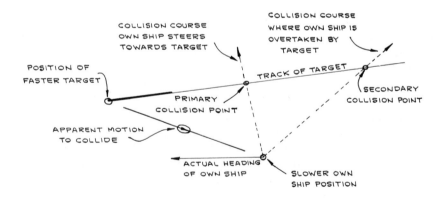

Figure 4.2   Dual points of collision

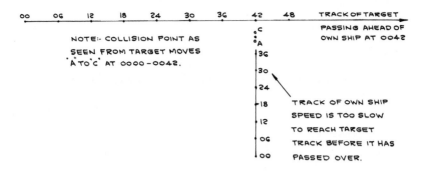

Figure 4.3 No points of collision

4.2 How do the collision points behave if own ship maintains speed?

4.2.1 In the initial collision case

It is important to realize that collision points exist, whether an actual collision threat exists or not. The only significance is that in the event of an actual collision threat the collision points are the same for both ships. Figure 4.4 shows how the collision points move in a collision situation, and how they will appear to the two ships involved.

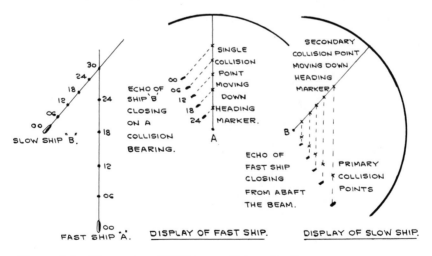

Figure 4.4 Movement of PPC in a collision situation

45

On the faster ship the single collision point appears on the heading marker and moves down, diminishing in range, as the collision approaches.

On the slower ship one of the two collision points will move down the heading marker while the other moves down a steady bearing.

4.2.2 In the non-collision case

In a non-collision case the collision point moves according to well defined rules, but it will never cross the heading marker.

In the case of the faster ship the movement depends on whether the fast ship will cross ahead of, or behind, the target ship. Figure 4.5 shows the two possible cases and typical track lines.

The case of the slower ship is more complicated because of the two collision points, and the possibility of no collision point existing. If the own ship course is to pass between the two collision points, they will pass down either side of own ship, generally shortening in range and then draw together under the stern of own ship. As they meet they become the one collision point before they finally disappear, as in Figure 4.6.

If own ship is steering to pass astern of the fast ship the collision points will draw together, form one point and then disappear. It is noticeable that the collision point more remote from the target ship, usually termed the secondary point, moves much faster than the point nearer to the target, the point termed the primary collision point.

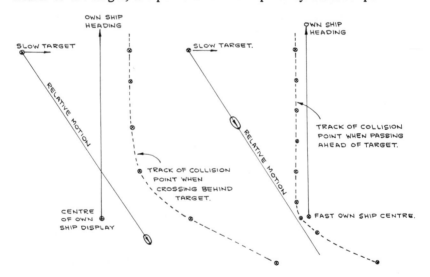

Figure 4.5 Track of collision point

46

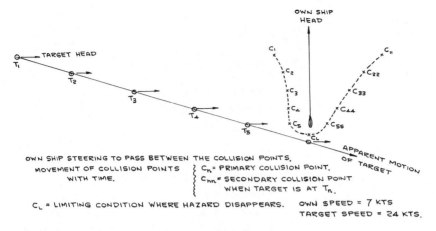

Figure 4.6 Movement of dual collision points

4.3 How does the collision point behave when own speed changes?

If the speed ratio is infinitely large, that is the target is stationary, then obviously the collision point is at the position of the target. If own ship maintains speed while the target begins to increase speed then the collision point will begin to move along the target track and when the target speed has increased to be the same as own ship, the secondary collision point will appear at infinity. Further increase of the target speed will move the primary and secondary collision points together; eventually own speed in comparison to target speed may be so slow that the two points will merge and disappear. This behaviour is shown in Figure 4.7.

4.4 How does the collision point move with change in target course?

Figure 4.8 shows how the collision point varies with changing aspect at differing speed ratios.

If the two ships are the same speed, the collision point moves on a locus which is the perpendicular bisector of the line joining the two ships. The greater the aspect the further away the collision point will be. Theoretically the limiting aspect in this case is 90° but in that case the collision point would be at infinity, and hence an aspect of some 85° plus is considered the practical limit.

47

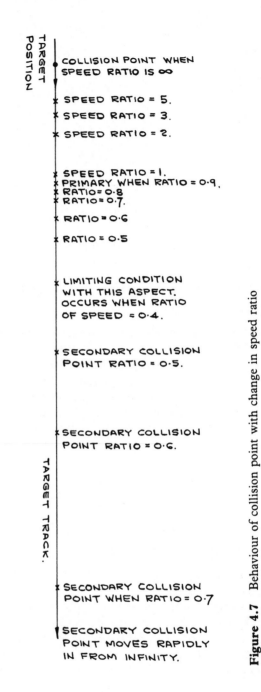

Figure 4.7 Behaviour of collision point with change in speed ratio

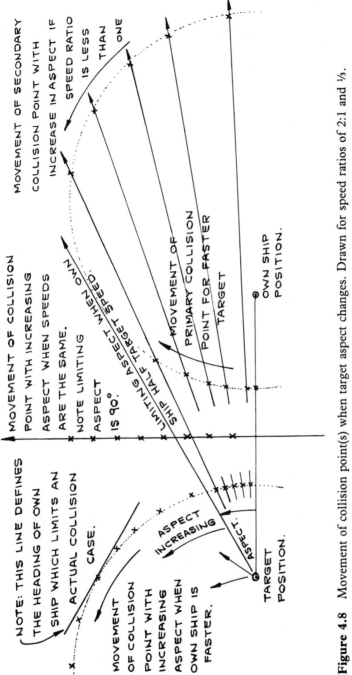

Figure 4.8 Movement of collision point(s) when target aspect changes. Drawn for speed ratios of 2:1 and ⅓.

4.4.1 For a slower own ship

When the own ship is slower than the target, the two collision points exist and they are seen to be on a circle whose centre and radius are dependent on the speed ratio; the circle is always on the 'own ship' side of the unity speed ratio locus. A limiting aspect can be defined which is also dependent on the speed ratio, a slower own ship means a target will have a smaller limiting aspect angle.

Aspects greater than the limit pose no hazard, since own ship can never catch up with the target.

4.4.2 For a faster own ship

When own ship is the faster, the circle of collision points lie on the target side of the equal speed locus. As the aspect increases the collision point moves further away from own ship, and there is no limiting aspect; collision is always possible.

Note: An interesting point in Figure 4.8 shows that the inverse of the idea of a limiting aspect to the slow own ship appears when own ship is faster; this is effectively a limiting course for own ship. If the actual heading is to the remote side of this line all collision points appear on the one bow. If own heading is inside this limiting direction, the collision point will move across the heading marker as the target changes aspect.

4.5 What is a Predicted Area of Danger?

The concept of collision points can be extended to incorporating other factors, such as:

(a) *inaccuracies* in data acquisition
(b) *the dimension* of the involved ships
(c) *the miss distance* which the mariner seeks to achieve

Whereas in the case of the collision point there is a course which intercepts the target's track at the given speed ratio, in the predicted area of danger there are generally two intersection points. One of these is where own ship will pass ahead of the target and the other where own ship passes astern of the target. The angle subtended by these two limiting courses will depend upon:

(a) *the speed ratio*
(b) *the position* of the target
(c) *the aspect* of the target

As shown in the case of the collision point, a faster ship must always generate a single cross-ahead and cross-astern position. A slower ship produces much more complex possibilities and, depending on the three variables noted above, may produce:

(a) two cross-ahead, two cross-astern points
(b) one cross-ahead, two cross-astern points
(c) two cross-astern points
(d) no hazard

In the case of the single or primary collision point, the position at which own ship will cross ahead of the target is always further from the target than the collision point, while the cross-astern is always nearer to the target.

In the case of a slower ship, where there is a secondary collision point, the second cross-ahead position is nearer to the target and the associated cross-astern position most remote from it, see Figure 4.9.

To indicate limits within 'the cross-ahead cross-astern' arc it is necessary to draw a bar parallel to the target's track and at the intended miss-distance closer to own ship position.

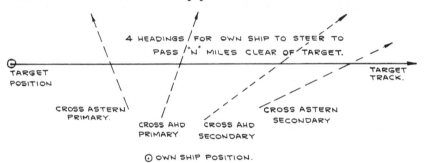
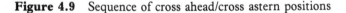

Figure 4.9 Sequence of cross ahead/cross astern positions

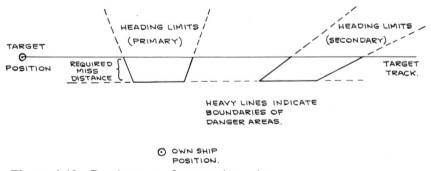
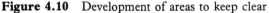

Figure 4.10 Development of areas to keep clear

The limits defined by the arc and the bar are such that if own ship should cross those limits, then it will be at a less distance than the desired miss-distance from the target. Figure 4.10 shows the generation of the two boundaries in the case of a slower own ship.

4.6   How is this concept adapted in practice?

In order to produce an acceptable system for practical operation, these limits are normally encapsulated by a symmetrical figure such as:

(a) *an ellipse*
(b) *a hexagon*

In the case of the ellipse the major axis is equal to the difference of the cross-ahead, cross-astern distances as measured from the target, and the minor axis is equal to twice the intended miss-distance. In the case of the hexagon, it is drawn from a rectangle and two isosceles triangles. The base of the triangle is always twice the miss-distance and the vertical height is one quarter of the distance E_1E_2 as shown in Figure 4.11. It should be noted that the collision point is not necessarily at the centre of either of the traditional figures.

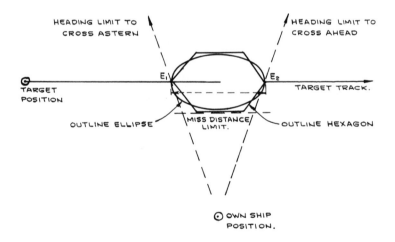

Figure 4.11   Acceptable symmetrical figures

In many cases the stylized figures do not follow the limits exactly, but any bias is on the safe side.

4.7 What causes a change in shape of the Predicted Area of Danger with change of miss-distance?

Due to the lack of symmetry in the geometry which generates the area, the cross-ahead and cross-astern positions do not move symmetrically about the collision point when the miss-distance is changed. The cross-ahead position usually moves more markedly than the cross-astern (*see* Figure 4.7, showing the movement of the two collision points, where the primary movement is much slower than the secondary). The overall result is an asymmetrical growth of the area with the cross-ahead position moving rapidly from the collision point.

4.8 How does the Predicted Area of Danger move?

4.8.1 In the collision case, or in the case of a less than desired miss-distance

As in the case of the collision point, when a danger area is violated by the heading marker the danger area will continue to move down the heading marker with the cross-ahead and cross-astern points on opposite bows. The shape of the danger area may change but it will never move off the heading marker. In the case of a slower ship where either of the two predicted areas is violated, the other will move in towards the target and eventually merge with the one on the heading marker.

In the limiting case where own ship heading marker just touches the limit of either predicted areas of danger, the limit will remain in contact with the heading marker, although the shape of the area may change considerably.

4.8.2 In the passing-clear case

In the non-collision case where the heading marker does not violate one of the danger areas, the areas themselves will move across the screen changing in shape and position. The movement will be very similar to that described for the collision point in Figure 4.5, depending on whether own ship is heading further ahead than the cross-ahead position or further astern than the cross-astern position. In the case of the dual areas of danger, although the movement will generally be the same as that shown for the dual collision point, a special case may arise when two danger areas may merge. This special case indicates the possibility of two cross-astern positions existing but no cross-ahead. It is also possible that cross-astern positions may exist, and an area of danger be drawn, which does not embrace an actual collision point.

4.8.3 Special cases

In some cases, for example, an end-on encounter, a cross-ahead and cross-astern position is not valid. In this context it is necessary to consider a pass to port and pass to starboard as defining the limits of the miss-distance. In the practical case, this results in the generation of a circle about the target's position.

Note: The justification for the Figures developed in this Section are more rigorously defined in Appendix IV.

Chapter 5

Sources of Error in Displayed Data

5.1 What error sources affect ARPA?

Errors which may influence decision making, and are present in data displayed on an ARPA screen are from three main areas. These error sources may be grouped as:

Group 1: *Errors which are generated in the radar installation* itself, the behaviour of the signals at the chosen frequency and the limitations of peripheral equipment such as logs, gyro and dedicated trackers.

Group 2: *Errors which may be due to inaccuracies during processing* of the radar data, inadequacies of the algorithms chosen and limits of accuracy accepted.

Group 3: *Errors in interpretation* of the displayed data.

5.2 What errors are generated in the radar installation? (Group 1)

Errors in the radar, gyro and log which feed data to the system will cause errors in the output data. Range and bearing errors which remain constant or nearly so during the encounter, e.g. a steady gyro error of a few degrees, will introduce error into the predicted course of other ships, but are unlikely to cause danger since all course data will be similarly affected – including own ship. The effect of errors on the predicted data depends on the kind of error, the situation, and the duration of the plot for which data is stored for processing and prediction. This time is typically in the range 1 to 3 minutes. In the following examples, the situation is assumed to be a near miss or a collision.

5.3 Glint

As a ship rolls, pitches and yaws, the apparent centre of its radar echo moves over the full ship's length; this is termed glint. Its distance from amidships is random with a standard deviation of one sixth length, i.e. for a 600 ft ship it is probable that the error does not exceed ± 100 ft. Since the beam of a ship is usually small in comparison to length, transverse glint is negligible. If the target ship's aspect is beam on, glint introduces random bearing errors.

5.4 Bearing errors

These cause false positions to be observed on each side of the relative track of the other ship, leading to errors in the observed relative track, and therefore in the predicted CPA and also in the displayed aspect of the other ship. Unfortunately, the greatest errors in displayed aspect occur in those cases where the real aspect is near end-on. Bearing errors may be due to:

5.4.1 Backlash

Backlash between rotating aerial and its azimuth transmitter. Air resistance on the rotating aerial will tend to maintain gear tooth contact, but bounce and reversed torque due to aerodynamic forces will break the contact and allow some backlash to occur.

5.4.2 Unstable platform or tilt

Ship motion causes the axis of rotation of the radar aerial to tilt. When the ship is heeled by B radians a bearing error of:

$$-(\tfrac{1}{2}.B^2.\sin \theta.\cos \theta) \text{ radians is produced,}$$

where θ is the bearing of the target ship off own ship's bow. This error is quadrantal, i.e. zero ahead, astern and abeam, rising to alternative + and − maxima at 45 and 135 degrees.

It will *not* be reversed by the opposite roll since B is squared. When the ship is rolling it has two components; a random variation between zero and a maximum, according to the value of B (i.e. the actual roll angle which happens to be present when the aerial is directed on the bearing), and a rise and fall of the maximum over periods of about one or two minutes with wave height variation. For A = 45 degrees and a roll of 7½ degrees towards or away from the other ship, the error is −0.25 degrees maximum.

5.4.3 Parallax due to roll of own ship

If the radar antenna is mounted at a height L above the roll axis of the ship, and the ship rolls to an angle B, the antenna moves sideways by L.sin B. The measured bearing of a target at a bearing of θ from the ship's head and at a range R will be in error by an angle e given by:

$$e = \frac{180.L.\sin B.\cos \theta}{\pi R} \text{ degrees}$$

(L and R must be in the same units)

This error varies sinusoidally with time and has a period equal to the roll period.

5.4.4 Asymmetrical antenna beam

The ARPA should take the bearing of the target as that of the centre of the echo. If the antenna beam is asymmetrical the apparent position of the echo may change with the echo strength. Errors due to this cause can become very large in some systems if the echo strength is sufficient for the close-in side lobe pattern of the antenna to become apparent. At least one system employs special techniques to eliminate this problem.

5.4.5 Azimuth quantization error

The antenna position must be converted to digital form before it can be used by the computer, for example, by using a shaft encoder. A 12-bit shaft encoder has a least-significant-bit equivalent to .09° [360°/4096] so that the restriction to 12 bits introduces a quantization error of 0.045°. The same error will arise if the computer truncates the input azimuth information to 12 bits. Antenna azimuth is often taken to a resolution of either 12 or 13 bits. Note: Since many gyro repeater systems are step-by-step, with a step size of $\frac{1}{6}$°, there is no real point in making the antenna encoder bit size very much smaller.

5.5 What errors occur in range measurements?

5.5.1 Distance change due to roll of own ship

If the radar antenna is mounted at a height L above the roll axis of the ship, and the ship rolls by an angle B, the antenna moves sideways by (L.sin B). For a target at a bearing θ from the ship's head, the measured range will be in error by a distance 'd' given by:

$$d = L.\sin B.\sin \theta$$

Pitch error is much less significant, but if roll and pitch occur together the effects add non-linearly and must be worked out separately.

5.5.2 Range quantization error

The range of a target must be converted into a digital number for the computer to use and it is likely that this will be done by measuring the range by counting techniques. A convenient clock rate is that corresponding to 0.01 nm steps, i.e. about 8 MHz.

Typical step function due to range and bearing quantizing are shown in Figure 5.1.

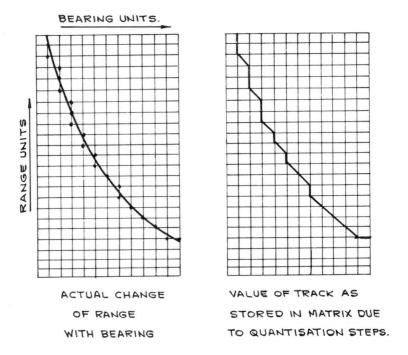

BEARING UNITS.

RANGE UNITS

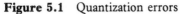

ACTUAL CHANGE
OF RANGE
WITH BEARING

VALUE OF TRACK AS
STORED IN MATRIX DUE
TO QUANTISATION STEPS.

Figure 5.1 Quantization errors
Note During some periods, it would appear that the target is on a collision course, i.e. steady bearing, although this is never the case in fact.

5.5.3 Pulse amplitude variation

The equipment will typically measure the range of an echo at the point at which the echo strength rises above a pre-set threshold. Because of the finite bandwidth of the radar receiver, the echo pulse will have a sloping leading edge and the measured range will vary with the pulse amplitude. For the pulse lengths commonly used on anti-collision range scales the receiver bandwidths are chosen for long range performance rather than for discrimination (typically 5 MHz), so it is likely that the leading edge slope will be nearly as long as the transmitted pulse.

The resulting apparent range variation will depend on some assumptions about echo amplitude variations, but are likely to be about 40 metres as the maximum.

5.6   What effects are gyro errors likely to have?

A gyro master unit mounted at a height above the roll axis of a ship is subject to transverse acceleration at each extremity of the roll. This includes a false vertical, and the pendulous unit tilts in its gymbals. This puts an error into the gyro output affecting all bearings; it has random and slowly varying components just as the radar tilt error. Observation at sea indicates that 0.25 deg is the error in many typical installations.

The gyro also has other errors. Long term errors (e.g. settling point $\pm$ 0.75 deg) are unimportant if they remain sensibly constant as they normally do, but random errors (e.g. settling point difference $\pm$ 0.2 deg) are significant.

5.6.1   Gyrocompass deck-plane (gimballing) errors

The true heading of a vessel is the angle between the vessel's fore and aft line and the meridian when measured in the horizontal plane. In several gyrocompass designs the sensing element has sufficient degrees of freedom to assume a north-south, horizontal attitude. However, the compass card may be constrained to the deck plane. In this case there can be a discrepancy between the compass card reading and the ship heading detected by the sensitive element.

5.6.2   Yaw motion produced by the coupling of roll and pitch motion

When a ship is rolling and pitching, these two motions interact to produce a resultant yawing motion. The motion can be resolved into horizontal and vertical components. The horizontal component is the yaw motion and is detected by the gyrocompass sensing element.

5.7 What effects are log errors likely to have?

An error in own ship's log will produce a vector error in every other ship's true speed and course. This will also result in an error in the displayed aspect of other ships; however this aspect error is minimum in all cases where the real aspect is end-on. A further effect will be to produce non-zero speed indications on all stationary targets (e.g. navigation marks and ships at anchor) being tracked. If this error is assumed not to exceed 0.4 knot, it will give rise to a positional error of some 17 yds in a plot time of 75 seconds.

5.8 Summation of errors – likely magnitude

Maximum bearing errors

Backlash	-0.2 to $+$	0.2 deg
Tilt (radar)	0 to $\pm$	0.25 deg
Tilt (gyro)	0 to $\pm$	0.25 deg
Gyro random	-0.2 to $+$	0.2 deg

Chapter 6

Errors in Displayed Data

6.1 What is target swop?

When two targets are close to each other, it is possible for the association of past and present echoes to be confused so that the processor is loaded with erroneous data.

The result is that the historical data on one target may be transferred to another, and the indicated relative (and true) track of that ship will be composed of part of the tracks of two different target motions. This target swop can occur with any type of tracker, but is least likely in those which use a diminishing gate size as the confidence in the track rises, and those which adopt longer periods of rate aiding. It is most likely to occur when two targets are close together for a comparatively long time and one target echo is much stronger than the other, see Figure 6.1.

6.2 What are likely track errors? (Group 2)

The apparent motion of a target is in fact rarely steady, and even steady motion will return positions which are randomly scattered about the actual track due to basic radar limitations. Quantization (step) errors in range and bearing, introduced by the translation of the basic radar information to the processor data base, further exacerbate these system errors. The only way that the tracker can deal with these is to use some form of smoothing over a period of time by applying more or less complicated mathematical filters. The aim of the filter is to give the best possible indication of the steady track, and at the same time detect real

61

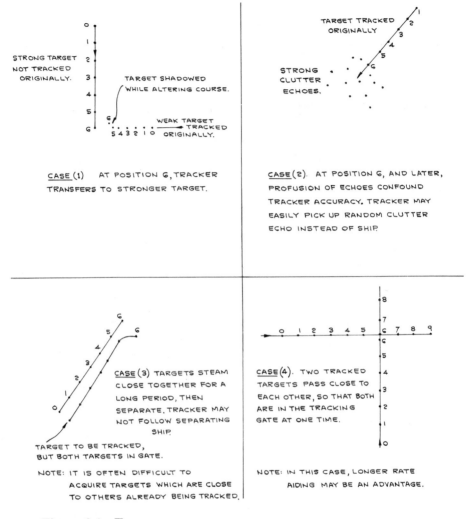

Figure 6.1 Target swop

changes quickly. (Clear weather observation of tracker performance compared to actual target behaviour is recommended). The success of the tracker can be judged from two criteria, the stability of the track shown on the display, and the speed with which an actual change in target real behaviour can be identified. Usually, when targets have a small relative motion, that is, they are steering approximately the same course as own ship, the quantization steps are more likely to give rise to 'jumps', apparent motion is least reliable and real changes are easily masked.

62

Similarly, when own ship is altering course the relative motion of the target will follow a curve. Trackers which are smoothing a relative track from relative positions will only obtain the mean track over the period. Resolution of the true motion may also be in error because the data extracted from gyro and log during the period of swing may be very different to the path which the ship's mass is actually following. This may be further aggravated if the immediate course and speed are applied to a track which has been smoothed over a discrete time period. Trackers which immediately convert the relative position of the target to the immediate cartesian True North values will obtain a more accurate appreciation of target true track, and most important, will obtain earlier warning of target alterations. However, if the apparent motion in this system is re-constituted by calculation, using gyro and log information, then it is important to bear in mind that, during the period of alteration, indication of the nearest approach quantities must be viewed with suspicion.

Watching the behaviour of a known stationary target during manoeuvres may give some indication of the accuracy of the overall performance of the system. Vectors appearing on stationary targets in the true mode may be subject to a combination of tide and errors. (see: influence of tide, page 77).

Note: When targets are at very short range the relative motion gives rise to very rapid bearing changes which may cause the tracker to give a target lost response.

6.3   What is the influence on vector presentation of errors due to incorrect data input?

6.3.1   Relative vectors

Systems which store the relative positions only will correctly indicate the relative track (apart from errors during own ship action) since the data portrayed is independent of any errors which exist in peripherals such as gyro or log. Systems which reconstitute relative motion from stored true data will also portray correct relative vectors. However, it is important to stress the need for the operator to continuously compare one data source against another to ensure that, in all cases, indications from true motion output and relative motion output sensibly agree.

6.3.2 True vectors

Systems which store true positions will display the true tracks correctly referenced to the heading marker, although an incorrect compass input may cause the picture to be slewed. In systems storing relative positions any errors in inputs from peripherals will introduce errors in the displayed information. In an actual collision case indication of collision from the true vectors will still be evident, but misassessment of the method of dealing with the situation may occur due to the erroneous aspect and speed. This is shown in figure 6.2. In a passing situation the problem may be sufficient to change appreciation of the time and distance of CPA and influence choice of manoeuvre.

Figure 6.3 shows such a situation, where a ship passing to starboard might induce a starboard manoeuvre.

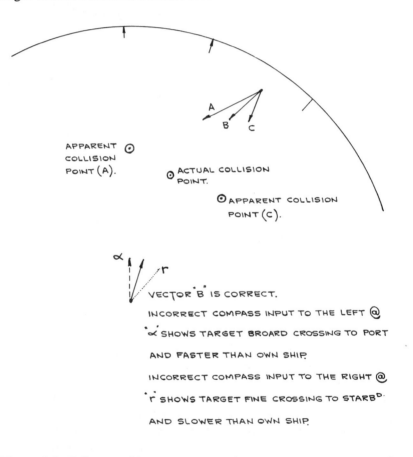

APPARENT COLLISION POINT (A).

ACTUAL COLLISION POINT.

APPARENT COLLISION POINT (C).

VECTOR B IS CORRECT.

INCORRECT COMPASS INPUT TO THE LEFT @

∝ SHOWS TARGET BROAD CROSSING TO PORT

AND FASTER THAN OWN SHIP.

INCORRECT COMPASS INPUT TO THE RIGHT @

r SHOWS TARGET FINE CROSSING TO STARB^D·

AND SLOWER THAN OWN SHIP.

Figure 6.2   Influence of incorrect compass input on true vector presentation

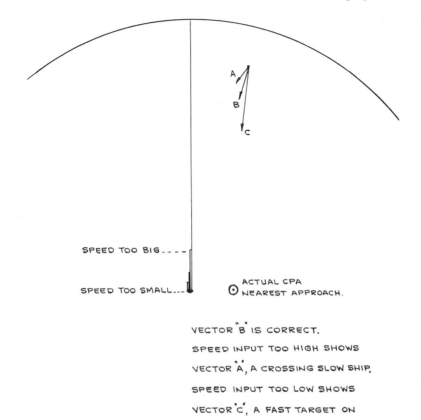

SPEED TOO BIG____

SPEED TOO SMALL___

ACTUAL CPA
⊙ NEAREST APPROACH.

VECTOR "B" IS CORRECT.

SPEED INPUT TOO HIGH SHOWS

VECTOR "A", A CROSSING SLOW SHIP.

SPEED INPUT TOO LOW SHOWS

VECTOR "C", A FAST TARGET ON

NEAR PARALLEL CROSSING.

Figure 6.3   Influence of incorrect speed input on true vector presentation

6.4   What is the influence on the position of the collision point due to incorrect data input?

6.4.1   Speed error

If incorrect speed input is given to a collision case situation, the collision point will still appear (correctly) on the heading marker but at an incorrect range and will move down the heading marker at an incorrect speed.

In the case where there is, in fact, a miss distance, the collision point will appear in the wrong position which may give rise to a misjudgement of the danger or urgency in a situation. Figure 6.4 shows how the collision point may be displaced due to speed error in two cases where the target is crossing ahead and crossing astern.

65

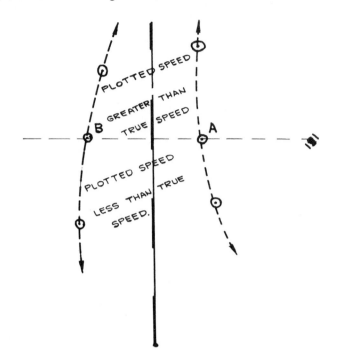

Figure 6.4   Influence of speed error on PPC
A = Target passing astern, correct speed used
B = Target passing ahead, correct speed used

6.4.2   Errors in course input

The behaviour of the collision point when an error in the course is input is too complex to allow definition of a pattern. If the error only occurs in the calculation, and does not appear in the position of the heading marker, the collision point could appear on the heading marker in a miss situation. More dangerously, a point could appear off the heading marker in a collision situation. When the same error appears in both heading marker and calculation, as might occur due to a gyro error, the collision case will always show the point on the heading marker. Similarly, if a miss distance exists the collision point will not be on the heading marker.

Chapter 7

Errors of Interpretation

7.1 Errors of interpretation (Group 3)

These errors are not within the system, but are those likely to be made by the operator through misunderstanding, inexperience or casual observation.

7.2 What errors are likely in the vector system?

In the case of vector systems the most common mistakes arise when the mode of presentation is mis-used; attempting to obtain nearest approach from true vectors or using the relative vector to indicate speed and course of the target. Some equipments fit spring-loaded switches to ensure the equipment is failed to one specific mode in an attempt to reduce the chance of this misinterpretation of data. Good practice is to check the graphic data against numeric data if it is available.

7.3 What errors are likely in the use of the danger area concept?

In the use of collision point and areas of danger the commonest mistakes arise when attempting to interpolate, or extrapolate, data from the display. Typical errors arise because of misconceptions of the following points:

(a) *the line joining the target to the collision point* is not a time-conscious vector and does not indicate speed

(b) *the collision point* gives no indication of miss distance

(c) *changes in collision point positions* do not indicate change in target true course or speed

(d) *the area of* danger does not change symmetrically with change in miss distance

(e) *the collision point* is not at the centre of the danger area

These points are illustrated in Figures 7.1 to 7.6.

It is always important to realize that the areas of danger generated on the screen apply only to own ship and the target, they do not always give warnings of a mutual threat between two targets. If two areas of danger overlap, it is reasonable to suppose that the two targets involved will also pass each other within the stated miss distance, but separate danger areas do not imply safe passing between targets. Two targets may eventually have a close passing although their danger areas, as applied to own ship, appear to be well separated.

7.3.1 Resumption of course

Where a 'chained' bearing cursor is available, and the chain divisions are an indication of time, care must be exercised in the measurement of time to resume course. As shown in Figure 7.7, the marker correctly indicates the time own ship will pass ahead and astern on the target track, but the time of the required miss distance occurring can not be determined.

7.4 Can afterglow be misleading?

It should be noted that because most ARPAs use a centre, or off-centred, fixed origin display, the *movement* of an echo is usually in its relative track. When used in the true mode, vectors and afterglow will not correlate. In the fewer displays which use a moving origin, and allow a relative vector display, the true afterglow will not match the relative vector.

7.5 How accurate is the presented data?

Over reliance on, and failure to appreciate inaccuracies in, presented data which has been derived from imperfect inputs should be avoided at all costs. It must always be borne in mind that a Vector/PAD/Alphanumeric readout is not absolutely accurate, just because it has been produced by a 'computer', no matter how many microprocessors it may

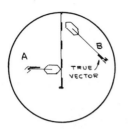

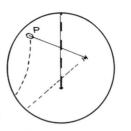

Figure 7.1 Errors of interpretation
Target A is faster and Target B is
slower than own ship, despite
appearances.
Note Vectors will show this

Figure 7.2 Errors of interpretation
Solid line shows track of PPC from
P. Apparent track of echo which will
occur is dotted.

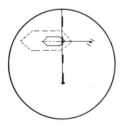

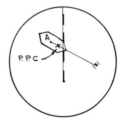

Figure 7.3 Errors of interpretation
Dotted PAD shown for two miles,
solid PAD shown for one mile.

Figure 7.4 Errors of interpretation
PPC is not at A, the hexagon centre.

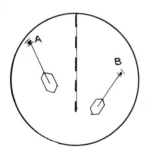

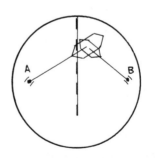

Figure 7.5 Errors of interpretation
Targets A and B will collide with
each other, although not apparent
from the display.

Figure 7.6 Errors of interpretation
Targets A and B will not collide
with each other although they may
pass within the miss distance.

69

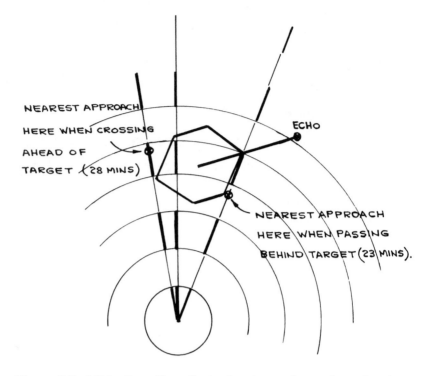

Figure 7.7 Misleading effect of using bearing marker to determine time to resume
Note There are 4 intervals to point of crossing ahead and 5 intervals to cross astern but nearest approach occurs 4.8 and 3.6 intervals respectively.

boast. An indication by ARPA that a target will pass 1 cable clear of own ship should not be regarded as justification for standing on into such a situation.

The errors given in table 3.8.3 of the IMCO Performance Standards for ARPA are quite typical and should always be allowed for.

7.6   Can targets be missed?

An automatic acquisition system may totally fail to detect and acquire a target of vital importance, for any one of a number of reasons. Similarly it may also drop or cancel a fading target. In the latter case the target may subsequently be re-acquired and present a course and speed which may indicate that the target has manoeuvred, when in fact, the track is new and has not yet established its long-term accuracy.

Chapter 8

Tests, Warnings and Alarms

8.1 Tests and warnings

In the case of conventional radar there was always a basic requirement to ensure that the radar controls were properly adjusted and that correct operation of the radar had been established by means of a performance check.

There is now a further requirement for the ARPA to provide warnings and alarms;

(a) *in the event of malfunction in the ARPA* section of the equipment; a means shall be provided whereby the observer can monitor the performance against a known solution. These are termed equipment warnings

(b) *to alert the observer where targets are likely to violate pre-set criteria.* These are termed operational warnings.

8.2 Equipment warnings

Most equipments incorporate self-diagnostic routines which monitor the correct operation of the various circuits. This check is repeated at regular intervals (which may range from once per hour to many times per second) or on request from the operator. In the event of a fault, a warning is given to the operator and, in most cases, some indication of the cause or location (e.g. printed circuit board No. 53) of the fault is also given. However, it must be appreciated that some faults cannot be detected internally e.g. a failure of certain elements in a numeric readout can cause an 8 to appear as a 3 and a 0 to appear as a 7 etc.

Another occurrence which will activate warnings is loss of a sensor input, such as log or gyro data missing. It is important here to note that the ARPA has no way of knowing what values to expect and so can only warn of their absence. To ensure that the processor which deals with the data has sufficient overall accuracy, four test scenarios are specified along with the tolerances within which the output data must fall. All ARPAs should be able to conform to this level of accuracy.

8.3 What are operational warnings?

8.3.1 Guard rings and zones

It is possible to specify an area in the vicinity of own ship e.g. at 10 miles a zone 5 cables in range which, if entered by a target, would activate an alarm. It is usual to have two zones, one of which may be at some pre-set range and the other variable. The target which has activated the alarm may be made to 'flash' or alternatively have a short bearing marker through it. In some equipments, the range and bearing of the target entering the zone is displayed in digital form. It is important to remember that a target which is detected for the first time *within* a guard ring will not activate the alarm, and that this warning system should not be regarded as an alternative to keeping a proper lookout. Each guard zone may be set for 360° coverage or for a specific sector of threat.

8.3.2 Close-quarter warnings

It is possible to specify a CPA and TCPA which will activate an alarm if *both* are violated. For example: if the CPA and TCPA controls are set to .5 nm and 30 mins respectively, and a target would have a CPA of less than .5 nm in less than 30 minutes, then the alarm will be activated. This will occur, even if the 'relative vector' has not been extended into the specified area. The displayed echo and vector of the target activating the alarm will be made to flash.

Where 'own ship' heading marker intersects a PAD, a warning will be activated, and will continue until such time as 'own ship' course is altered to clear the PAD.

8.3.3 Target lost

Consider a target which is being tracked but, for one of a number of reasons, does not return a detectable response on one scan. It is required that the tracker should continue to search for the echo in an area where it might be expected (see rate aiding) for up to five successive scans. If after this searching the echo is not re-acquired, the 'target lost' warning would be activated and the last observed position of the echo marked on the screen.

8.3.4 Wrong request

Where an operator feeds in wrong data or data in an unacceptable form, e.g. course 370°, an alarm and indicator will be activated, and will continue until the invalid data is deleted or overwritten.

8.4 What additional alarms may be provided?

Additional alarms to warn the operator may be fitted. These are activated:

(a) *when a tracked target is found* to have made a significant manoeuvre
(b) *while at anchor*, if a specified target(s) shows movement. This can be used to indicate if 'own ship' is dragging or swinging, or if another vessel is getting under way
(c) *when it is time to manoeuvre* as previously assessed under the 'trial' facilities
(d) *when all the tracking channels are being used.* This will warn the operator to inspect the untracked targets for potential dangers, and transfer tracking from a less important target.

Audio and visual alarms are normally provided and while it is usually possible to silence the audio alarm, the visual alarm will continue until the cause for the alarm or warning has been removed.

Chapter 9

Trial Manoeuvre Facilities and History

9.1 What are trial manoeuvre facilities?

With the availability of computer assistance, the problem of predicting the effect of a manoeuvre prior to its implementation by own ship are much simplified. While it is relatively easy to mentally visualize the outcome of a manoeuvre where two ships are involved, in high density traffic it becomes very difficult. Particularly with large ships and limited sea room, it is necessary to plan and update the whole collision avoidance strategy in light of the continually changing radar scene as quickly as possible.

It is important to bear in mind, while planning, that:

(a) *own ship may temporarily need to be on a 'collision course'* with more distant vessels, i.e. collisions may require sequential avoidance since it is unlikely that a single manoeuvre will resolve all the problems.

(b) *extrapolation of the 'present' situation* using the trial manoeuvre facility with current course and speed can provide valuable information on which of the 'other' vessels in the vicinity may have to manoeuvre in order to avoid collisions between each other. Obvious avoiding manoeuvres may appear and should be watched for.

(c) *constraints imposed by navigation* may dictate the manoeuvre of 'other' vessels. This should be taken into account when planning strategy and watched for when carrying out the plan and assessing its effectiveness.

Approved ARPAs are required to possess the facility for simulating such a trial manoeuvre. Unfortunately, a variety of different methods of providing this facility have been devised by manufacturers.

9.2   How are trial manoeuvres displayed?

In all cases, the proposed course and speed are fed in. In some equipments it is also possible to feed in a time delay, intended rudder angle or ship handling characteristics in various states of loading.

Various methods of displaying such simulation are at present available:

9.2.1   A vector display (see Figure 9.1)

With the information relating to the proposed manoeuvres, the true or relative vectors which would result from such a manoeuvre would be displayed. When combined with the ability to adjust the vector length, this mode can give a clear presentation of potential close-quarter situations between other vessels as well as with own ship.

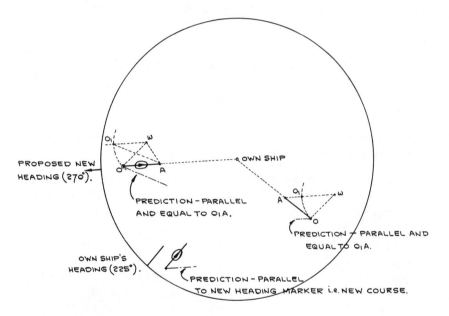

Figure 9.1   Forecast facilities—static display

9.2.2   A dynamic display

In this case all tracked targets, as well as own ship, are made to move at some multiple of normal speed (30 times real speed). Own ship will move at the proposed speed in the proposed direction with 'own ship' handling characteristics and delay (if any) being taken into account.

In this prediction mode potential close-quarter situations (and the likelihood of evasive manoeuvres) between other vessels may also be readily apparent, but the forward predicted situation cannot be 'frozen' for detailed analysis.

9.3   What indicates that the trial manoeuvre function is in operation?

When the equipment is being used in the simulation mode, it is required that means shall be provided for ensuring that it cannot be mistaken for either of the 'live' modes. Typical means of preventing confusion are:

(a) *A spring loaded switch* to invoke the forecast mode.
(b) *A legend such as 'SIM'* is made to appear on the screen.

Note:
(i) While trial manoeuvres are in progress on the display, the computer continues its normal task of tracking all marked targets.
(ii) Where PADs are provided, the necessary alterations of course should be apparent continuously but a 'trial speed' facility is provided.

9.4   What is a history display?

It is required that an approved ARPA should be able to display, on request, at least four equally time-spaced past positions of any targets being tracked over a period of at least eight minutes. This enables an observer to check whether a particular target has manoeuvred in the recent past, possibly while the observer was temporarily away from the display on other bridge duties.

It should be appreciated that history of a target's behaviour may give some indication of its future intention. Changes in motion may be for navigational purposes but it very often means a collision avoidance action being undertaken and a resumption of previous conditions can be watched for.

History tracks may be shown in either true or relative mode, changes in direction of motion may not necessarily mean changes in course when the relative mode is used. Checks for changes in target behaviour should be made by observation of the true information.

Uneven tracks of targets, or apparent instability of motion may be taken to indicate that tracking of that target is less precise than it might be, and the displayed data should be treated with caution.

Chapter 10

Ground Lock and Navigation Lines

10.1 What is sea stabilized mode?

When ARPAs are used in the True Tracks mode, data relating to own ship's motion is fed in from log and compass. If it is accepted that there are no sensor errors and that the log is feeding in speed through the water (not from a bottom locked doppler log) then the displayed true tracks would be *sea stabilized* (see Figure 10.1). Vectors would therefore indicate the true tracks through the water of other vessels and would thus indicate the visual aspects of other vessels, irrespective of tide experienced.

IT IS ESSENTIAL WHEN THE ARPA IS USED IN THE TRUE TRACK ANTI—COLLISION ROLE, THAT IT IS ONLY USED IN THE SEA STABILIZED MODE.

10.2 How does the coastline appear in sea stabilized mode?

In the sea stabilized true tracks mode, if the ship is experiencing tide, any displayed 'land' echoes will be seen to drift. This may prove to be a nuisance when using a 'Map' display.

10.3 What is land locked, or ground stabilized mode?

Coastline drift can be prevented by feeding in tidal corrections or better still – on some ARPAs – using the 'auto drift' or 'echo reference' facility. In these cases the display will be locked to some fixed target and

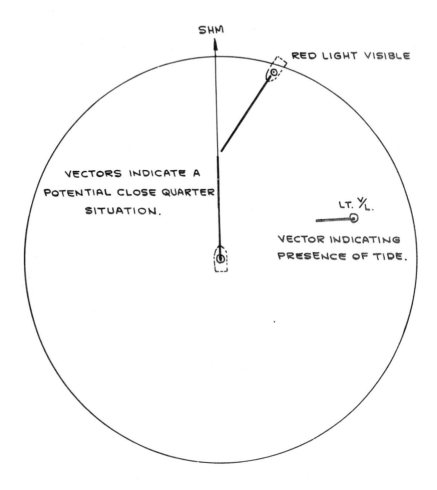

Figure 10.1 Sea stabilized display

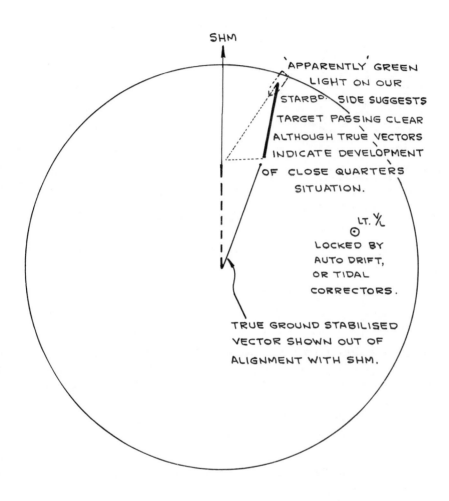

SHM

'APPARENTLY' GREEN
LIGHT ON OUR
STARB⁰ SIDE SUGGESTS
TARGET PASSING CLEAR
ALTHOUGH TRUE VECTORS
INDICATE DEVELOPMENT
OF CLOSE QUARTERS
SITUATION.

LT. ⅟

LOCKED BY
AUTO DRIFT,
OR TIDAL
CORRECTORS.

TRUE GROUND STABILISED
VECTOR SHOWN OUT OF
ALIGNMENT WITH SHM.

Figure 10.2   Ground stabilized display

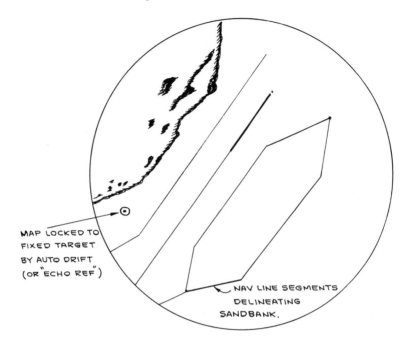

MAP LOCKED TO
FIXED TARGET
BY AUTO DRIFT
(OR "ECHO REF")

NAV LINE SEGMENTS
DELINEATING
SANDBANK.

Figure 10.3 Map presentation

becomes *ground stabilized*. The displayed vectors will then indicate target's true tracks over the ground. Because of the potentially misleading effect of data relating to tracked vessel's aspects (see Figure 10.2), this mode should not be used when assessing collision risk or planning avoidance strategy.

10.4 What are 'Nav Lines' or Map Presentations?

Most ARPAs now have available some means by which lines can be made to appear on the display. They are adjustable in position, orientation and length and can be used to delineate navigational limits in channels, for parallel indexing techniques (see Figure 10.4), to map poorly responding coastlines, etc. The number of available segments or elements is of the order of 15 to 20, so that only crude maps can be drawn (see Figure 10.3) but it is possible to store up to fifteen such maps (even when the radar is switched off) in some equipment.

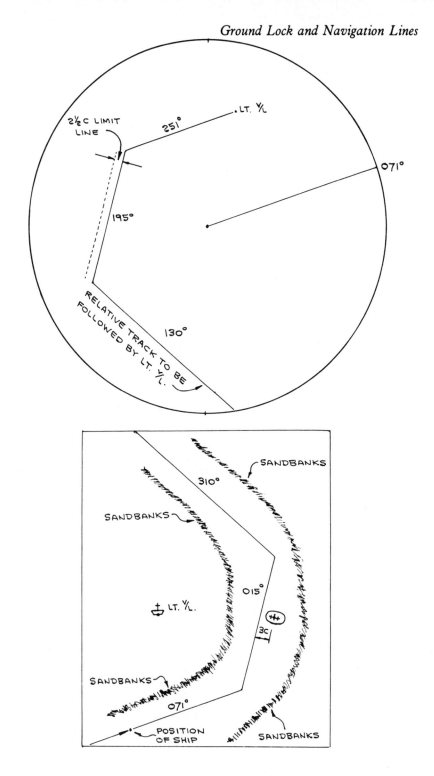

Figure 10.4   Use of Nav Lines for parallel indexing

Chapter 11

The Collision Regulations – The Relevance of ARPA

11.1 Introduction

A notable feature of the 1972 Collision Regulations is that many specific references to the use of radar are made in the body of the rules. While some of these originated in the Annex to the previous regulations, some of the references are new. Some relevant extracts from the regulations are given in Appendix V.

When considering the application of ARPA to collision avoidance, it is particularly relevant that the term 'radar plotting or *equivalent systematic observation* of detected objects' (Rule 7b) appears for the first time. Although the rules do not define this term, consideration of the various instructions and cautions given in the specific references to radar make it possible to deduce a procedure which should enable competent personnel, with the aid of ARPA, to comply with the letter and spirit of the rules. In this respect ARPA should be seen as readily providing data which would otherwise have to be obtained by lengthy and tedious manual extraction.

11.2 Obligations to use radar equipment

Any discretion as to whether or not to make use of operational radar, if fitted, which may have existed under the previous rules appears to have been removed by Rule 7b. This section specifies that proper use shall be made of such equipment, and the well-established warning about scanty information is now embodied in this rule. Although Rule 5 (Lookout) does not specifically mention radar, there seems little doubt that it is embraced in the term 'all means available'; for its ability to detect targets and in its role as a source of information allows an observer to make a more complete appraisal of the situation. It would also appear that there is an implied requirement to use the equipment in clear weather where it can augment or clarify the visual scene, e.g. in high density traffic, especially at night.

11.3 Safe speed

Rule 6, in listing the factors to be considered when determining a safe speed, devotes a complete section to those factors which can be determined by the use of radar. It is important to realise that the factors listed extend beyond the context of basic radar into that of collision avoidance systems.

11.4 Development of a close-quarters situation

In cases where a target is detected by radar alone, Rule 19d places a specific obligation on the observer to determine if a close-quarters situation is developing. To comply with this requirement the target should be acquired, tracked and its predicted range at CPA extracted. On most systems both the distance and predicted time of CPA can be made available in alpha-numeric form if the target is designated. (The interval to CPA will assume particular importance if a close-quarters situation is developing). The numeric output should be cross-checked by reference to any graphical presentation of the data; exactly how the graphic data is presented will depend on the type of system employed.

At this juncture, it is the duty of the observer to decide if the predicted range at CPA constitutes a close-quarters situation; the range at which a close-quarters situation is considered to arise will depend on factors such as:

(a) *the geographical position*
(b) *the handling capability* of the ship
(c) *the density of the traffic*

11.5 Avoidance manoeuvre

If it is decided that a close-quarters situation is developing, the observer must take action to resolve the situation in ample time, subject to the recommendations laid down in Rule 19d(i) and (ii). Any avoiding manoeuvre not based on a knowledge of the true motion of all relevant targets would be unseamanlike (see DoT Notice M983, 3.6) and could certainly attract criticism as an assumption based on scanty radar information.

When ARPA is being used, the true motion data may be obtained in alpha-numeric form and cross-checked by interpretation of the graphical presentation, but since these are often generated from the same database a good comparison should not be taken as an assumption of accuracy of the information. It is important to remember that the success of action by own ship may be influenced by recent changes in the target's true motion. It is thus essential to check, by using the history presentation, whether such changes have occurred.

11.6   Forecast

Rule 8d requires that action taken to avoid collision with another vessel shall be such as to result in a safe passing distance. This rule applies to all states of visibility, but in restricted visibility the safe passing distance may have to be greater than that which could be tolerated in clear weather.

When employing ARPA passing distances may, and should, be forecast by employing the trial manoeuvre facility appropriate to the system in use. Such forecast will also enable the observer to check that the manoeuvre will not result in a close-quarters situation with other vessels (Rule 8c) and can often indicate where targets may manoeuvre to avoid each other.

It is essential to remember that a further criterion for an acceptable manoeuvre is that it must be readily apparent to another vessel observing by radar (Rule 8b). In making this judgement one must consider own ship's speed and the speed of target vessels. Further it must be appreciated that, even if the target vessel is equipped with radar, the plotting facilities may be very basic. Hence the rate at which he can extract data, and thus become aware of changes, may be slow and his ability to identify small changes may be particularly limited. (Thus Rule 8b gives warning about a succession of small alterations).

Finally, the possibility that a target vessel may not be plotting – or may not even have operational radar – must be continually kept in mind.

11.7   Effectiveness of manoeuvre

Rule 8d requires that the effectiveness of the avoiding action be checked until the other vessel is finally past and clear. This requirement can be satisfied by monitoring:

(a) *the true motion* of all relevant targets (to ensure early detection and identification of target manoeuvres)

(b) *the relative motion* of all relevant targets to check for fulfilment of the forecast nearest approach.

11.8 Resumption

When the target is finally past and clear, the decision must be made to resume course and/or speed.

As in the case of the avoiding manoeuvre, prior to resuming, the trial manoeuvre facility should be employed to verify that a safe passing distance will be achieved with respect to all relevant targets. It should be remembered that the resumption will be most obvious to other observing vessels if it is performed as a single manoeuvre. The common practice of resuming course in steps by 'following a target round' will make it difficult for other observing ships to identify the manoeuvre positively and could be considered to be in contravention of the spirit, if not the letter, of Rule 8b.

11.9 Conclusion

In any potential collision situation, particularly in restricted visibility, the interpretation of displayed radar information facilitates the determination and execution of action to avoid close-quarters situations. Traditionally this was achieved by decision-making based on data extracted by manual plotting. ARPA should be seen as a device which extracts and presents such data. It thus reduces the workload on the observer by carrying out routine tasks and allows him more time to carry out decision-making on the basis of the data supplied.

The ability of the equipment to carry out routine tracking and computation in no way relieves the observer of the need to understand fully the principles of radar plotting, or to be capable of applying such principles to a practical encounter, such as might be the case in the event of an equipment failure.

It is imperative that the observer is capable of interpreting and evaluating the data presented by the system. Equally essential is the ability to detect any circumstances in which equipment is producing data which is inconsistent with the manner in which a situation is developing as observed, say, from the raw radar.

In general it is vital that the observer understands the limitation of the system in use and hence is aware of the dangers of exclusive reliance on the data produced by ARPA.

In particular, the implicit reliance on the validity of the prediction of small non-zero passing distances should be avoided. The CPA errors, tabulated in sections 3.8.2 and 3.8.3 of the Performance Standards for ARPA, clearly indicate that predicted passing distances of less than 1 mile should be treated with the utmost caution.

SUMMARY

1 USE ARPA (AND RADAR) IF IT CAN BE OF THE SLIGHTEST ASSISTANCE
   (a) CLEAR WEATHER WITH HEAVY TRAFFIC
   (b) PROPER LOOKOUT

2 DOES RISK OF COLLISION EXIST, AND/OR IS A CLOSE-QUARTER SITUATION DEVELOPING?
   (a) ANALYSE DISPLAYED INFORMATION
     • RELATIVE TRACKS CPA & TCPA
     • TRUE TRACKS – CO. & SPD
     • HISTORY OF PAST POSITION

3 DETERMINE BEST MANOEUVRE
   (a) TRIAL FACILITY
     • WILL TARGETS PASS SAFELY?
     • WILL TARGETS HAVE TO MANOEUVRE?
   (b) AVOID A/C TO PORT, OR TURNING TOWARD CONVERGING OVERTAKING SHIPS

4 MANOEUVRE
   (a) BOLD – AVOID SUCCESSION OF SMALL A/C
   (b) LARGE – OBVIOUS TO OTHERS
   (c) IN GOOD TIME – TIME TO CORRECT
   (d) GOOD SEAMANSHIP

5 ENSURE ITS EFFECTIVENESS
   (a) CONTINUOUS CHECK OF RELATIVE MOTION AND HISTORY
   (b) WATCH FOR CHANGES IN TRUE MOTION OF TARGETS

6 RESUME
   (a) ARE TARGETS CLEAR?
   (b) TRIAL – CHECK OTHER V/Ls FOR LIKELY MANOEUVRES
   (c) MAKE ONE MANOEUVRE

Proposed Carriage Requirements for Automatic Radar Plotting Aids

IMCO – Safety of Navigation – Regulation 12

1 An automatic radar plotting aid of a type approved by the administration and conforming to performance standards not inferior to those adopted by the Organization (IMCO) shall be fitted on:

1 each ship of 10,000 grt or more, the keel of which is laid or is at a similar stage of construction on or after 1 September 1984;
2 existing tankers of 40,000 grt or more from 1 January 1985;
3 existing tankers of 10,000 grt or more from 1 January 1986;
4 other existing ships of 40,000 grt or more from 1 September 1986;
5 other existing ships of 20,000 grt or more from 1 September 1987;
6 other existing ships of 15,000 grt or more from 1 September 1988;

except that automatic radar plotting aids fitted prior to 1 January 1984 which do not fully conform to the performance standards recommended by the Organization may at the discretion of the Administration, be retained until 1 January 1991.

2 The Administration may exempt ships from this requirement, in areas where it considers it unreasonable or unnecessary for such equipment to be carried, or when the ship will be taken permanently out of service within 2 years of the appropriate implementation date.

Except that automatic radar plotting aids fitted prior to 1 January 1984 which do not fully conform to the performance standards adopted by the Organisation may be retained until 1 January 1991, at the discretion of the Administration.

The Administration may exempt from this requirement, in areas where it considers it unreasonable or unnecessary for such equipment to be carried, or when the ship will be permanently taken out of service within 2 years after the appropriate implementation date, provided that it is acceptable to the Governments of the States to be visited by the ship.

Appendix I(b)

Performance Standards for Automatic Radar Plotting Aids

(IMCO – Res. A422(XI) – Annex)

1 Introduction

1.1 The Automatic Radar Plotting Aids (ARPA) should, in order to improve the standard of collision avoidance at sea:

1 reduce the work-load of observers by enabling them to automatically obtain information so that they can perform as well with multiple targets as they can by manually plotting a single target;
2 provide continuous, accurate and rapid situation evaluation.

1.2 In addition to the General Requirements for Electronic Navigational Aids, the ARPA should comply with the following minimum performance standards.

2 Definitions

2.1 Definitions of terms used in these performance standards are given in Appendix I(c).

3 Performance Standards

3.1 Detection

3.1.1 Where a separate facility is provided for detection of targets, other than by the radar observer, it should have a performance not inferior to that which could be obtained by the use of the radar display.

3.2 Acquisition

3.2.1 Target acquisition may be manual or automatic. However, there should always be a facility to provide for manual acquisition and cancellation. ARPAs with automatic acquisition should have a facility to suppress acquisition in certain areas. On any range scale where acquisition is suppressed over a certain area, the area of acquisition should be indicated on the display.

3.2.2 Automatic or manual acquisition should have a performance not inferior to that which could be obtained by the user of the radar display.

3.3 Tracking

3.3.1 The ARPA should be able to automatically track, process and simultaneously display and continuously update the information on at least:

1 20 targets, if automatic acquisition is provided, whether automatically or manually acquired;
2 10 targets, if only manual acquisition is provided.

3.3.2 If automatic acquisition is provided, description of the criteria of selection of targets for tracking should be provided to the user. If the ARPA does not track all targets visible on the display, targets which are being tracked should be clearly indicated on the display. The reliability of tracking should not be less than that obtainable using manual recordings of successive target position obtained from the radar display.

3.3.3 Provided the target is not subject to target swop, the ARPA should continue to track an acquired target which is clearly distinguishable on the display for 5 out of 10 consecutive scans.

3.3.4 The possibility of tracking errors, including target swop, should be minimized by ARPA design. A qualitative description of the effects of error sources on the automatic tracking and corresponding errors should be provided to the user, including the effects of low signal to noise and low signal to clutter ratios caused by sea returns, rain, snow, low clouds and non-synchronous emissions.

3.3.5 The ARPA should be able to display on request at least 4 equally time-spaced past positions of any targets being tracked over a period of at least 8 minutes.

3.4 Display

3.4.1 The display may be a separate or integral part of the ship's radar. However, the ARPA display should include all the data required to be provided by a radar display in accordance with the performance standards for navigational radar equipment.

3.4.2 The design should be such that any malfunction of ARPA parts producing data additional to information to be produced by the radar, should not affect the integrity of the basic radar presentation.

3.4.3 The size of the display on which ARPA information is presented should have an effective display diameter of at least 340 mm.

3.4.4 The ARPA facilities should be available on at least the following range scales:

1 12 or 16 miles;
2 3 or 4 miles.

3.4.5 There should be a positive indication of the range scale in use.

3.4.6 The ARPA should be capable of operating with a relative motion display with 'north-up' and either 'head-up' or 'course-up' azimuth stabilization. In addition, the ARPA may also provide for a true motion display. If true motion is provided, the operator should be able to select for his display either true or relative motion. There should be a positive indication of the display mode and orientation in use.

3.4.7 The course and speed information generated by the ARPA for acquired targets should be displayed in a vector or graphic form which clearly indicates the target's predicted motion. In this regard:

1 ARPA presenting predicted information in vector form only should have the option of both true and relative vectors;
2 an ARPA which is capable of presenting target course and speed information in graphic form, should also, on request, provide the target's true and/or relative vector;
3 vectors displayed should be either time adjustable or have a fixed time-scale;
4 a positive indication of the time-scale of the vector in use should be given.

3.4.8 The ARPA information should not obscure radar information in such a manner as to degrade the process of detecting targets. The display of ARPA data should be under the control of the radar observer. It should be possible to cancel the display of unwanted ARPA data.

3.4.9 Means should be provided to adjust independently the brilliance of the ARPA data and radar data, including complete elimination of the ARPA data.

3.4.10 The method of presentation should ensure that the ARPA data is clearly visible in general to more than one observer in the conditions of light normally experienced on the bridge of a ship by day and by night. Screening may be provided to shade the display from sunlight but not to the extent that it will impair the observers' ability to maintain a proper lookout. Facilities to adjust the brightness should be provided.

3.4.11 Provisions should be made to obtain quickly the range and bearing of any object which appears on the ARPA display.

3.4.12 When a target appears on the radar display and, in the case of automatic acquisition, enters within the acquisition area chosen by the observer or, in the case of manual acquisition, has been acquired by the observer, the ARPA should present in a period of not more than one minute an indication of the target's motion trend and display within three minutes the target's predicted motion in accordance with paragraphs 3.4.7, 3.6, 3.8.2 and 3.8.3.

3.4.13 After changing range scales on which the ARPA facilities are available or re-setting the display, full plotting information should be displayed within a period of time not exceeding four scans.

3.5 Operational warnings

3.5.1 The ARPA should have the capability to warn the observer with a visual and/or audible signal of any distinguishable target which closes to a range or transits a zone chosen by the observer. The target causing the warning should be clearly indicated on the display.

3.5.2 The ARPA should have the capability to warn the observer with a visual and/or audible signal of any tracked target which is predicted to close to within a minimum range and time chosen by the observer. The target causing the warning should be clearly indicated on the display.

3.5.3 The ARPA should clearly indicate if a tracked target is lost, other than out of range, and the target's last tracked position should be clearly indicated on the display.

3.5.4 It should be possible to activate or de-activate the operational warnings.

3.6 Data requirements

3.6.1 At the request of the observer the following information should be immediately available from the ARPA in *alphanumeric* form in regard to any tracked target:

1 present range to the target;
2 present bearing of the target;
3 predicted target range at the closest point of approach (CPA);
4 predicted time of CPA (TCPA);
5 calculated true course of target;
6 calculated true speed of target.

3.7 Trial manoeuvre

3.7.1 The ARPA should be capable of simulating the effect on all tracked targets of an own ship manoeuvre without interrupting the updating of target information. The simulation should be initiated by the depression either of a spring-loaded switch, or of a function key, with a positive identification on the display.

3.8 Accuracy

3.8.1 The ARPA should provide accuracies not less than those given in paragraphs 3.8.2 and 3.8.3 for the four scenarios defined in Appendix I(d). With the sensor errors specified in Appendix I(e), the values given relate to the best possible manual plotting performance under environmental conditions of plus and minus ten degrees of roll.

3.8.2 An ARPA should present within one minute of steady state tracking the relative motion trend of a target with the following accuracy values (95% probability values):

Data

Scenario	Relative Course (degrees)	Relative Speed (knots)	CPA (nm)
1	11	2.8	1.3
2	7	0.6	
3	14	2.2	1.8
4	15	1.5	2.0

3.8.3 An ARPA should present within three minutes of steady state tracking the motion of a target with the following accuracy values (95% probability values):

Data

Scenario	Relative Course (degrees)	Relative Speed (knots)	CPA (nm)	TCPA (mins)	True Course (degrees)	True Speed (knots)
1	3.0	0.8	0.5	1.0	7.4	1.2
2	2.3	0.3			2.8	0.8
3	4.4	0.9	0.7	1.0	3.3	1.0
4	4.6	0.8	0.7	1.0	2.6	1.2

3.8.4 When a tracked target, or own ship, has completed a manoeuvre, the system should present in a period of not more than one minute an indication of the target's motion trend, and display within three minutes the target's predicted motion, in accordance with paragraphs 3.4.7, 3.6, 3.8.2 and 3.8.3.

3.8.5 The ARPA should be designed in such a manner that under the most favourable conditions of own ship motion the error contribution from the ARPA should remain insignificant compared to the errors associated with the input sensors, for the scenarios of Appendix I(d).

3.9 Connections with other equipment

3.9.1 The ARPA should not degrade the performance of any equipment providing sensor inputs. The connection of the ARPA to any other equipment should not degrade the performance of that equipment.

3.10 Performance tests and warnings

3.10.1 The ARPA should provide suitable warnings of ARPA malfunction to enable the observer to monitor the proper operation of the system. Additionally test programmes should be available so that the overall performance of ARPA can be assessed periodically against a known solution.

Appendix I(c)

Definitions of Terms to be Used Only in Connection with ARPA Performance Standards

Relative course	The direction of motion of a target related to own ship as deduced from a number of measurements of its range and bearing on the radar. Expressed as an angular distance from North.
Relative speed	The speed of a target related to own ship, as deduced from a number of measurements of its range and bearing on the radar.
True course	The apparent heading of a target obtained by the vectorial combination of the target's relative motion and ship's own motion.* Expressed as an angular distance from North.
True speed	The speed combination of a target obtained by the vectorial combination of its relative motion and own ship's motion.*
Bearing	The direction of one terrestrial point from another. Expressed as an angular distance from North.

---

*For the purpose of these definitions there is no need to distinguish between sea or ground stabilization.

Relative motion display	The position of own ship on such a display remains fixed.
True motion display	The position of own ship on such a display moves in accordance with its own motion.
Azimuth stabilization	Own ship's compass information is fed to the display so that echoes of targets on the display will not be caused to smear by changes of own ship's heading.
/North-up	The line connecting the centre with the top of the display is North.
/Head-up	The line connecting the centre with the top of the display is own ship's heading.
/Course-up	An intended course can be set to the line connecting the centre with the top of the display.
Heading	The direction in which the bows of a vessel are pointing. Expressed as an angular distance from North.
Target's predicted motion	The indication on the display of a linear extrapolation into the future of a target's motion, based on measurements of the target's range and bearing on the radar in the recent past.
Target's motion trend	An early indication of the target's predicted motion.
Radar plotting	The whole process of target detection, tracking, calculation of parameters and display of information.
Detection	The recognition of the presence of a target.
Acquisition	The selection of those targets requiring a tracking procedure and the initiation of their tracking.

97

Tracking   The process of observing the sequential changes in the position of a target, to establish its motion.

Display   The plan position presentation of ARPA data with radar data.

Manual   An activity which a radar observer performs, possibly with assistance from a machine.

Automatic   An activity which is performed wholly by a machine.

Appendix I(d)

Operational Scenarios

For each of the following scenarios, predictions are made at the target position defined after previously tracking for the appropriate time of one or three minutes:

Scenario 1

Own ship course	000°
Own ship speed	10 kt
Target range	8 nm
Bearing of target	000°
Relative course of target	180°
Relative speed of target	20 kt

Scenario 2

Own ship course	000°
Own ship speed	10 kt
Target range	1 nm
Bearing of target	000°
Relative course of target	090°
Relative speed of target	10 kt

Scenario 3

Own ship course	000°
Own ship speed	5 kt
Target range	8 nm
Bearing of target	045°
Relative course of target	225°
Relative speed of target	20 kt

Scenario 4

Own ship course	000°

Own ship speed	25 kt
Target range	8 nm
Bearing of target	045°
Relative course of target	225°
Relative speed of target	20 kt

Appendix I(e)

Sensor Errors

The accuracy figures quoted in paragraph 3.8 are based upon the following sensor errors and are appropriate to equipment complying with performance standards for shipborne navigational equipment.*

Note: σ means 'standard deviation'.

Radar

Target glint (Scintillation) (for 200 metres length target)
Along length of target = 30 m (normal distribution)
Across beam of target = 1 m (normal distribution)

Roll-pitch bearing The bearing error will peak in each of the four quadrants around own ship for targets on relative bearings of 045°, 135°, 225° and 315° and will be zero at relative bearings of 0°, 90°, 180° and 270°. This error has a sinusoidal variation at twice the roll frequency. For a 10° roll the mean error is

0.22° with a 0.22° peak sine wave superimposed.

Beam shape – assumed normal distribution giving bearing error with $\sigma = 0.05$.

---

*In calculations leading to the accuracy figures quoted in paragraph 3.8, these sensor error sources and magnitudes were used. They were arrived at during discussions with national government agencies and equipment manufacturers and are appropriate to equipments complying with the Organization's draft performance standards for radar equipment.

Independent studies carried out by national government agencies and equipment manufacturers have resulted in similar accuracies, where comparisons were made.

Pulse shape – assumed normal distribution giving range error with $\sigma = $ 20 metres.

Antenna backlash – assumed rectangular distribution giving bearing error $\pm$ 0.5° maximum.

Quantizations

Bearing – rectangular distribution $\pm$ 0.01° maximum.
Range – rectangular distribution $\pm$ 0.01 nm maximum.
  Bearing encoder assumed to be running from a remote synchro giving bearing errors with a normal distribution $\sigma = 0.03°$.

Gyro compass

Calibration error 0.5°
Normal distribution about this with $\sigma = 0.12°$

Log

Calibration error 0.5 kt
Normal distribution about this, $3\,\sigma = 0.2$ kt.

Appendix II

Proposed Minimum Requirements for Training in the Use of Automatic Radar Plotting Aids (ARPA) (relevant to Chapter II of the International Convention on Standards of Training, Certification and Watchkeeping for Seafarers, 1978)

1 Every master, chief mate and officer in charge of a navigational watch on a ship fitted with an automatic radar plotting aid shall have completed an approved course of training in the use of automatic radar plotting aids.
2 The course shall include the material set out below.

MINIMUM TRAINING REQUIREMENT IN THE OPERATIONAL USE OF AUTOMATIC RADAR PLOTTING AIDS (ARPA)

1 In addition to the minimum knowledge of radar equipment, masters, chief mates and officers in charge of a navigational watch on ships carrying ARPA shall be trained in the fundamentals and operation of ARPA equipment and the interpretation and analysis of information obtained from this equipment.
2 The training shall ensure that the master, chief mate and officers in charge of a navigational watch have:

(a) Knowledge of:
    (i) the possible risks of over-reliance on ARPA;

(ii) the principal types of ARPA systems and their display characteristics;

(iii) the IMCO performance standards for ARPA;

(iv) factors affecting system performance and accuracy;

(v) tracking capabilities and limitations of ARPA;

(vi) processing delays.

(b) Knowledge and ability to demonstrate in conjunction with the use of an ARPA simulator or other effective means approved by the administration:

(i) setting up and maintaining ARPA displays;

(ii) when and how to use the operational warnings, their benefits and limitations;

(iii) system operational tests;

(iv) when and how to obtain information in both relative and true motion modes of display, including:

- identification of critical echoes,
- use of exclusion areas in the automatic acquisition mode,
- speed and direction of targets relative movement,
- time to and predicted range at targets closest point of approach,
- course and speed of targets,
- detecting course and speed changes of targets and the limitations of such information,
- effect of changes in own ship's course or speed or both,
- operation of the trial manoeuvre;

(v) manual and automatic acquisition of targets, their respective limitations;

(vi) when and how to use true and relative vectors and typical graphic representation of target information and danger areas;

(vii) when and how to use information on past positions of targets being tracked;

(viii) application of the International Regulations for Preventing Collisions at Sea.

RECOMMENDED TRAINING PROGRAMME IN THE OPERATIONAL USE OF AUTOMATIC RADAR PLOTTING AIDS (ARPA)

1   General

(a) In addition to the minimum knowledge of radar equipment, masters, chief mates and officers in charge of a navigational watch on ships carrying ARPA should be capable of demonstrating a knowledge of the fundamentals and operation of ARPA equipment and the interpretation and analysis of information obtained from this equipment.

(b) Training facilities should include the use of simulators or other effective means capable of demonstrating the capabilities, limitations and possible errors of ARPA.

(c) The simulation facilities mentioned above should provide a capability such that trainees undergo a series of real-time exercises where the displayed radar information, at the choice of the trainee or as required by the instructor, is either in the ARPA format or in the basic radar format. Such flexibility of presentation will enable realistic exercises to be undertaken, providing for each group of trainees the widest range of displayed information available to the user and thus consolidating his ability to use effectively either basic radar or ARPA system.

(d) The ARPA training programme should include all items listed in paragraphs 3 and 4 below.

2 Training programme development

Where ARPA training is provided as part of the general training requirements, masters, chief mates and officers in charge of a navigational watch should understand the factors involved in decision-making based on the information supplied by ARPA in association with other navigational data inputs, having a similar appreciation of the operational aspects and of system errors of modern electronic navigational systems. This training should be progressive in nature commensurate with responsibilities of the individual and the certificate issued.

3 Theory and demonstration

3.1 The possible risks of over-reliance on ARPA

Appreciation that ARPA is only a navigational aid and that its limitations, including those of its sensors, make over-reliance on ARPA dangerous, in particular for keeping a look-out; the need to comply at all times with the basic principles and operational guidance for officers in charge of a navigational watch.

3.2 The principal types of ARPA systems and their display characteristics

Knowledge of the principal types of ARPA systems in use; their various display characteristics and an understanding of when to use ground or sea stabilised modes and north up, course up or head up presentations.

3.3 IMCO performance standards for ARPA

An appreciation of the IMCO performance standards for ARPA, in particular the standards relating to accuracy.

3.4 Factors affecting system performance and accuracy

(a) Knowledge of ARPA sensor input performance parameters – radar, compass and speed inputs; effects of sensor malfunction on the accuracy of ARPA data.
(b) Effects of the limitations of radar range and bearing discrimination and accuracy; the limitations of compass and speed input accuracies on the accuracy of ARPA data.
(c) Knowledge of factors which influence vector accuracy.

3.5 Tracking capabilities and limitations

(a) Knowledge of the criteria for the selection of targets by automatic acquisition.
(b) Factors leading to the correct choice of targets for manual acquisition.
(c) Effects on tracking of 'lost' targets and target fading.
(d) Circumstances causing 'target swop' and its effects on displayed data.

3.6 Processing delays

The delays inherent in the display of processed ARPA information, particularly on acquisition and re-acquisition or when a tracked target manoeuvres.

3.7 When and how to use the operational warnings, their benefits and limitations

Appreciation of the uses, benefits and limitations of ARPA operational warnings; correct setting, where applicable, to avoid spurious interference.

3.8   System operational tests

(a) Methods of testing for malfunctions of ARPA systems, including functional self-testing.
(b) Precautions to be taken after a malfunction occurs.

3.9   Manual and automatic acquisition of targets and their respective limitations

Knowledge of the limits imposed on both types of acquisition in multi-target scenarios, effects on acquisition of target fading and target swop.

3.10   When and how to use true and relative vectors and typical graphic representation of target information and danger areas

(a) Thorough knowledge of true and relative vectors; derivation of targets' true courses and speeds.
(b) Threat assessment; derivation of predicted closest point of approach and predicted time to closest point of approach from forward extrapolation of vectors, the use of graphic representation of danger areas.
(c) Effects of alterations of courses and/or speeds of own ship and/or targets on predicted closest point of approach and predicted time to closest point of approach and danger areas.
(d) Effects of incorrect vectors and danger areas.
(e) Benefit of switching between true and relative vectors.

3.11   When and how to use information on past position of targets being tracked

Knowledge of the derivation of past positions of targets being tracked, recognition of historic data as a means of indicating recent manoeuvring of targets and as a method of checking the validity of the ARPA's tracking.

4   Practice

4.1   Setting up and maintaining displays

(a) The correct starting procedure to obtain the optimum display of ARPA information.
(b) Choice of display presentation; stabilized relative motion displays and true motion displays.

(c) Correct adjustment of all variable radar display controls for optimum display of data.

(d) Selection, as appropriate, of required speed input to ARPA.

(e) Selection of ARPA plotting controls, manual/automatic acquisition, vector/graphic display of data.

(f) Selection of the time scale of vectors/graphics.

(g) Use of exclusion areas when automatic acquisition is employed by ARPA.

(h) Performance checks of radar, compass, speed input sensors and ARPA.

4.2 System operational tests

System checks and determining data accuracy of ARPA including the trial manoeuvre facility by checking against basic radar plot.

4.3 When and how to obtain information from ARPA display

Demonstrate ability to obtain information in both relative and true motion modes of display, including:

• identification of critical echoes,

• use of exclusion areas in the automatic acquisition mode,

• speed and direction of targets relative movement,

• time to and predicted range at target's closest point of approach,

• course and speed of targets,

• detecting course and speed changes of targets and the limitations of such information,

• effect of changes in own ship's course or speed or both,

• operation of the trial manoeuvre.

4.4 Application of the International Regulations for Preventing Collisions at Sea

Analysis of potential collision situations from displayed information, determination and execution of action to avoid close-quarter situations in accordance with International Regulations for Prevention of Collisions at Sea.

Appendix III

Plotting Revision

Since most of what the computer has to do is the same as the plotter would normally do, it is felt timely to include some revision of basic plotting.

If you plot the examples given, do not worry if your answers differ slightly from those given in the text, so long as your answer closely resembles the diagram as plotted.

AIII.1   The basic plot

With 'own ship' steering 000°(T) at a speed of 12 knots, an echo is observed as follows:

  0923 echo bears 037°(T) range 9.5 nm
  0929 echo bears 036°(T) range 8.0 nm
  0935 echo bears 034°(T) range 6.5 nm

Determine: CPA, TCPA, course, speed and aspect of the target.

1 Draw in own ship's Heading Line on the plotting sheet.
2 Plot the first position of the target. Label it 'O'. Insert the time.
3 Lay off a line to represent the apparent motion of a stationary target, from 'O' opposite to the direction of the heading line.
4 Plot at least one intermediate position to ascertain that the apparent motion is not changing (insert time).
5 Plot the final position, with time, and label it 'A'. See that the apparent motion is consistent in direction and rate. Join OA and label thus:

6 Produce the apparent motion line OA to find the closest point of approach (CPA) and time to CPA (TCPA).

7 The 'plotting interval' is the time between the readings of range and bearing for 'O' and 'A'. Find the distance that Own Ship has steamed in this time, and plot the position in which you would expect to find a stationary target 'W' at the end of the plotting interval. (OW is derived from Own Ship's speed and course reversed).

8 If 'A' and 'W' coincide, then the target is stationary. If they do not, then the line 'W' to 'A' represents the proper motion of the target in the plotting interval.

9 'Aspect' is measured between the 'line of sight' and the WA direction as in the diagram.

ANSWER: 1.0 nm in 25 mins., course 276°(T), speed 10½ knots. Aspect Red 62° – see Figure AIII.1.

AIII.2   The plot where only targets alter

Your own ship is steering 310°(T) at a speed of 12 knots. Echoes are observed as follows:

Time	Echo 'A'	Echo 'B'
0923 brg	270°(T) Range 9.0 nm	brg 347°(T) Range 9.4 nm
0929 brg	270°(T) Range 7.5 nm	brg 346°(T) Range 8.0 nm
0935 brg	270°(T) Range 6.0 nm	brg 344°(T) Range 6.5 nm

(a) Determine the CPA and TCPA, course and speed of each target.

Plotting is continued:

0941 brg	270°(T) Range 4.5 nm	brg 341°(T) Range 5.0 nm
0947 brg	253°(T) Range 2.8 nm	brg 350°(T) Range 4.0 nm
0953 brg	206°(T) Range 1.7 nm	brg 004°(T) Range 3.1 nm

(b) Say what action each target has taken (Your own vessel maintains course and speed throughout).

The apparent motion line OA, which is produced to find nearest approach time and distance, depends on four factors:

(a) Own ship's course
(b) Own ship's speed
(c) Target's course
(d) Target's speed

If any of these factors change, then the Apparent Motion line will also change. Changes in Own Ship's motion can be predicted (but in this

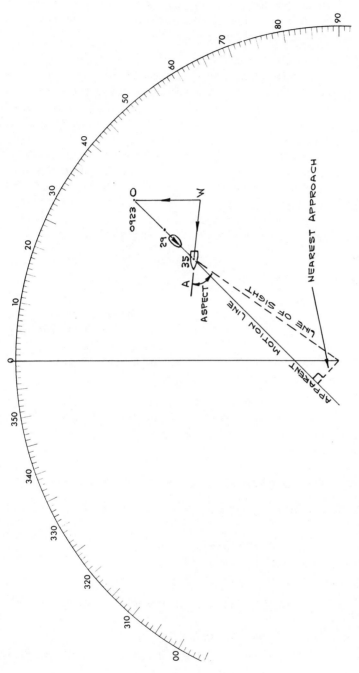

Figure AIII.1 The basic plot

example remain unchanged) and the new apparent motion pre-calculated, but to determine the change in the target's proper motion, the following procedure should be adopted.

1 Own ship's 'Heading Line' should have been drawn in.
2 The original Apparent Motion Line should have been plotted, OAW triangle constructed, and proper motion of the target ascertained.
3 Continue to plot the apparent motion of the target, making sure that the rate and direction are as predicted. (This is best accomplished by frequent plotting).
4 *As soon as a change in the target's apparent motion is noticed*, plot even more frequently, and when a new *steady* apparent motion is established, choose a position on it and label it O_1 and insert the time.
5 Plot the new apparent motion in the same way as if it were a new target. Predict the new CPA and TCPA – also target's course and speed, from which any changes from the previous proper motion may be noted.

ANSWER: (see Figure AIII.2)

Target A	*Target B*
(a) CPA = Collision in 24 mins (i.e. 0959) course 040°(T) speed 10 knots	(a) 1.0 nm in 25 mins (i.e. 1000) course 226°(T) speed 10½ knots
(b) target has altered course only. (50° to starboard)	(b) target has stopped

Note: Echo 'B' is similar to that used in example 1 and although in this case, the bearings are re-orientated, the plot and answer should be identical with that in Example AIII.1.

AIII.3 When own ship alters course only

With own ship steering 000°(T) at a speed of 12 knots an echo is observed as follows:

0923 echo bears 037°(T) range 9.5 nm
0929 echo bears 036°(T) range 8.0 nm
0935 echo bears 034°(T) range 6.5 nm

at 0935 it is intended to alter course (assume instantaneous) 60° to starboard

(a) Predict the new CPA and TCPA.
(b) Predict the new CPA and TCPA if the manoeuvre was delayed until 0941.

(c) Predict the bearing and range of the echo at 0953, if the manoeuvre was made at 0941.

1 The original OAW triangle should have been constructed.
2 With compasses at 'W' draw an arc of radius equal to WO.
3 Draw from 'W' a line, at an angle to port or starboard of WO equal to the proposed alteration of course.
4 Label the position where this line cuts the arc, O_1.
5 Join O_1A. (This now represents the new apparent motion in direction and rate).
6 Draw in the *ship's new heading line*, and expunge old heading line.
7 Until the alteration takes place, the target will continue to move down the original apparent motion line. Predict the position of the target at the time at which it is proposed to alter Own Ship's course, label O_2.
8 Draw O_2A_2 parallel with and equal to O_1A and produce, if necessary, to find the new nearest approach.

Note: The prediction is based on the assumption that the target will maintain course and speed. If the target makes an alteration, the prediction will not come true, and the target will have to be re-plotted.

ANSWER: (see Figure AIII.3)

(a) 4.4 nm in 13 mins from alteration, i.e. 0948.
(b) 3.6 nm in 10 mins from alteration, i.e. 0951.
(c) bearing 335°(T), range 3.6 nm

AIII.4   When own ship alters speed only

With own ship steering 000°(T) at a speed of 12 knots, an echo is observed as follows:

0923 echo bears 037°(T) range 9.5 nm
0929 echo bears 036°(T) range 8.0 nm
0935 echo bears 034°(T) range 6.5 nm

at 0935 it is intended to reduce speed to 3 knots (assume instantaneous).

(a) Predict the new CPA and TCPA.
(b) Predict the new CPA and TCPA if manoeuvre is delayed until 0941.
(c) Predict the bearing and range of the echo at 0953 if the manoeuvre is delayed until 0941.

1 The original OAW triangle should have been constructed.

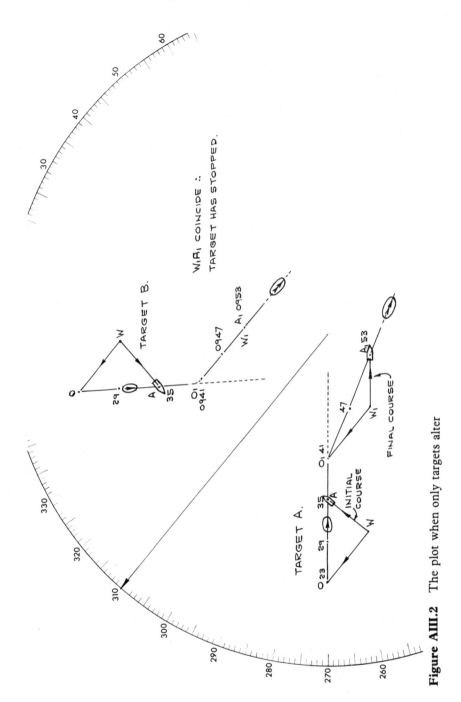

Figure AIII.2   The plot when only targets alter

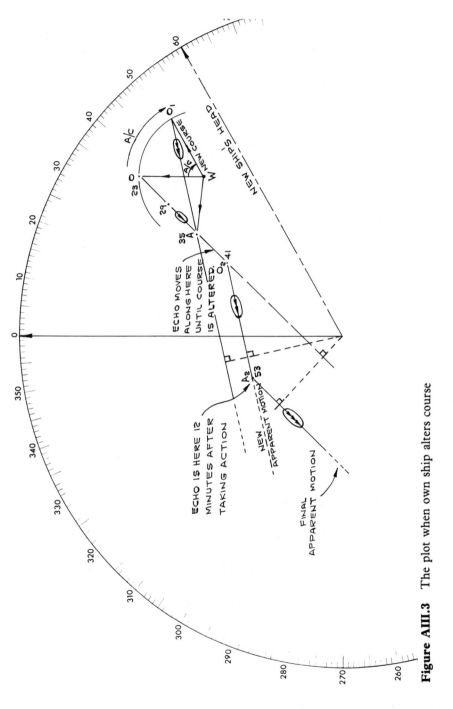

Figure AIII.3 The plot when own ship alters course

2 From 'W' measure off in the direction WO, the distance own ship will steam in the plotting interval, at the new speed, label the point O_1. Join O_1A (this represents the direction and rate of the new apparent motion).

3 Until the proposed alteration takes place, the target should continue to move along the old apparent motion line. Predict the position of the target at the time at which it is proposed to alter Own Ship's speed. Label it O_2.

4 Lay off O_2A_2 parallel with and equal to O_1A and produce it if necessary, to find the new nearest approach.

Note: This prediction is based on the assumption that the target will maintain its course and speed, and that own change in speed is instantaneous.

ANSWER: (see Figure AIII.4)
(a) 6.7 nm in 24½ mins from alteration in speed (i.e. 0959½)
(b) 5.8 nm in 19 mins from alteration in speed (i.e. 1000)
(c) bearing 007°(T), range 5.0 nm

Note: It should be appreciated that when changing speed, especially in large vessels, it can take a considerable time to achieve the required new speed and because of this, the new apparent motion line O_2A_2 will be much nearer own ship than predicted in the example above. In ARPAs the ship's handling characteristics (usually for only one condition of loading) may be allowed for when using the trial manoeuvre facility.

AIII.5   When own ship resumes course and/or speed

In taking action to avoid a close quarters situation, the action should be *substantial enough, and held for long enough*, to show up in the plotting of radar observers on other vessels.

To determine the effect of resuming own ship's course and/or speed:

1 The original OAW triangle should have been constructed.

2 The apparent motion after altering Own Ship's course and/or speed should have been predicted, and provided the target has kept its course and speed, the target will follow the new apparent motion line (O_2A_2).

3 If the target still maintains its course and speed and own ship

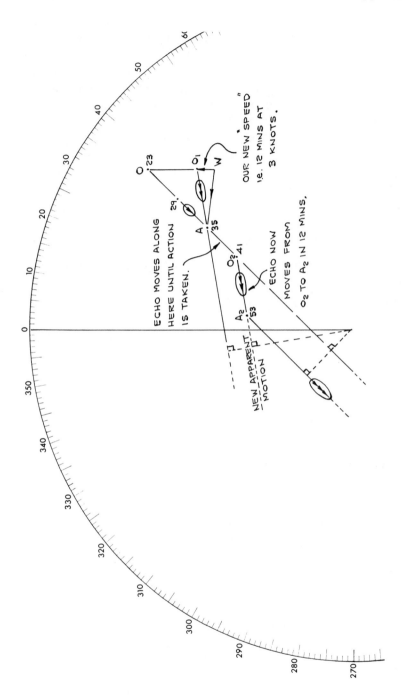

Figure AIII.4 The plot when own ship reduces speed

117

resumes her original course and/or speed, the direction and rate of apparent motion will be the same as in the original OAW triangle. Thus, from the target's predicted position at the time of resumption, draw O_3A_3 parallel with and equal to OA, and produce if necessary to find nearest approach after resumption.

4 If it is required to find the time to resume, so that the nearest approach will not be less than – for example – 4 miles, then lay off a line parallel to OA and at a tangent to the 4-mile range ring.

The point at which this apparent motion line crosses the second apparent motion line should be labelled O_3. Predict the time at which the target will reach O_3 – this will be the time to resume.

Note: When predicting the movement of a target along an apparent motion line, be sure to use the appropriate *rate* of apparent motion.

Using the examples in AIII.3 and AIII.4, predict in each case the final CPA if the original course and speed respectively is resumed at 0953.

ANSWER: (see Figure AIII.3) if course resumed at 0935, CPA = 3.5 nm

(see Figure AIII.4) if speed is (instantaneously) resumed at 0953, CPA = 2.4 nm

Appendix IVa

The PPC (Possible Point of Collision)

To find the PPC(s): (see Figure AIV.1).

1 Plot the target and produce the basic triangle.
2 Join the 'own ship' position to 'now position', CA, and produce beyond A.
3 With compasses at W and radius WO scribe the arc to cut CA produced at O_1 (or if own ship is the slower, i.e. WO < WA, at O_1 and O_2).
4 Join WO_1 (and WO_2).
5 Draw CP_1 parallel to WO_1 to cut WA produced at P_1 (and CP_2 parallel to WO_2 to cut WA produced at P_2).
6 P_1 (and P_2) is the PPC.

For a clearer appreciation of the determination of the PPC the following plot should be drawn out full size and to scale.

With own shp steering 000°(T) at 10 knots, an echo is plotted as follows:

0923 echo bears 037°(T) at 10.3 nm
0929 echo bears 036°(T) at 8.5 nm
0935 echo bears 034°(T) at 6.7 nm

Determine the bearing and range of the PPCs.

ANSWER: P_1 = 337°(T) at 4.4 nm (see Figure AIV.2)
$\quad\quad\quad\quad$ P_2 = 270°(T) at 18.0 nm

120

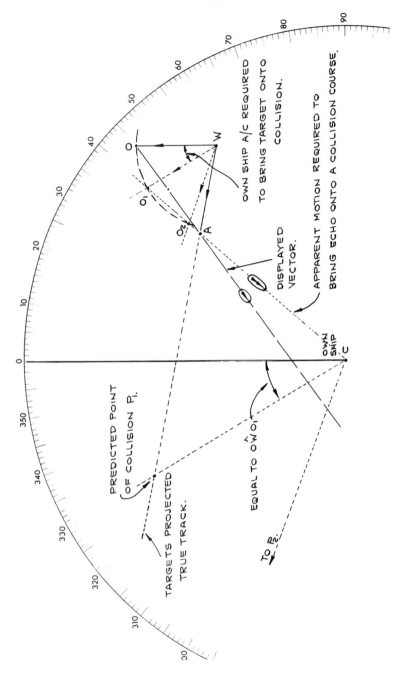

Figure AIV.1   The PPC

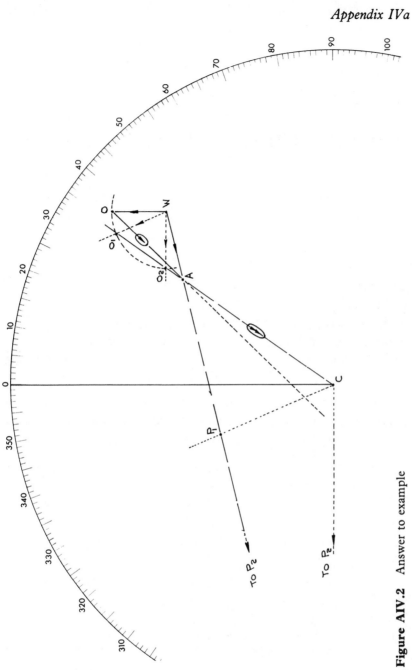

Figure AIV.2 Answer to example

Appendix IVb

The PAD (Predicted Area of Danger)

To construct the PAD (see Figure AIV.3).

1 Plot the target and produce the basic triangle.
2 Draw lines AT_1 and AT_2 from the now position (A), tangential to a circle of radius 'required CPA' and produce beyond A.
3 With compasses at W and radius WO scribe an arc to cut T_1A produced and T_2A produced at O_1 and O_2 respectively.
4 Join WO_1 and WO_2. (These represent the limiting courses to steer to clear the target by the 'required CPA.')
5 Draw CE_1 and CE_2 parallel to WO_1 and WO_2 respectively, to cut WA produced in E_1 and E_2 respectively.
6 At the mid-point of E_1E_2 draw the perpendicular to E_1E_2 and extend in both directions. In each direction, mark off the 'required CPA' and label E_3 and E_4.
7 Draw in the ellipse passing through the points, E_1, E_3, E_2, E_4; *or* draw in the hexagon as indicated in Figure AIV.4.

For a clearer appreciation of the construction of the PAD, the following plot should be drawn out full size and to scale:

With own ship steering 000°(T) at 12 knots, an echo is plotted as follows:

    1000 echo bears 045°(T) at 10.0 nm
    1006 echo bears 045°(T) at 8.5 nm
    1012 echo bears 045°(T) at 7.0 nm

Plot the target and draw in the hexagonal PAD for a 2.0 nm clearing.

ANSWER: See Figure AIV.4.

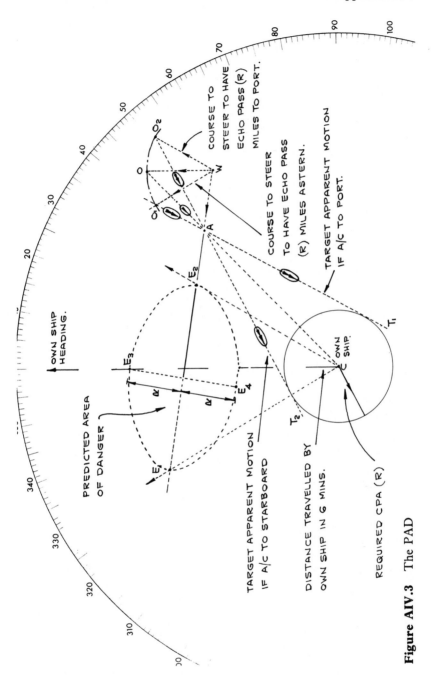

Figure AIV.3 The PAD

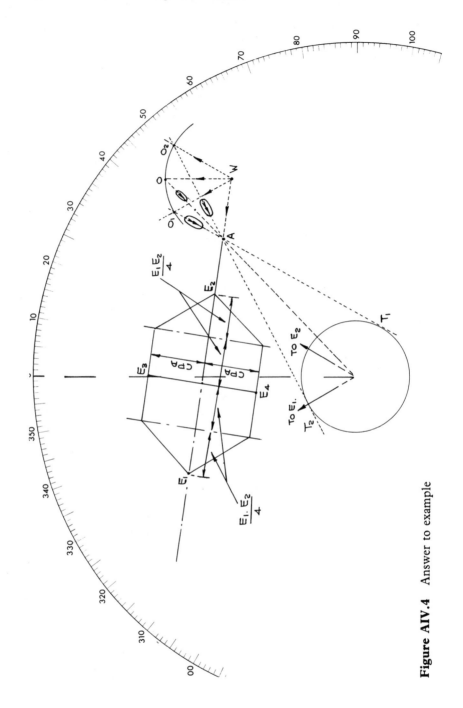

Figure AIV.4 Answer to example

Relevant Extracts from the International Regulations for Preventing Collisions at Sea (1972)

Rule 5

Look-out

Every vessel shall at all times maintain a proper look-out by sight and hearing as well as by all available means appropriate in the prevailing circumstances and conditions so as to make a full appraisal of the situation and of the risk of collision.

Rule 6

Safe speed

Every vessel shall at all times proceed at a safe speed so that she can take proper and effective action to avoid collision and be stopped within a distance appropriate to the prevailing circumstances and conditions.

In determining a safe speed the following factors shall be among those taken into account:

(a) By all vessels:
    (i) the state of visibility;
    (ii) the traffic density including concentrations of fishing vessels or any other vessels;
    (iii) the manoeuvrability of the vessel with special reference to stopping distance and turning ability in the prevailing conditions;

     (iv) at night the presence of background light such as from shore lights or from back scatter of her own lights;

     (v) the state of wind, sea and current, and the proximity of navigational hazards;

     (vi) the draught in relation to the available depth of water.

(b) Additionally, by vessels with operational radar:

     (i) the characteristics, efficiency and limitations of the radar equipment;

     (ii) any constraints imposed by the radar range scale in use;

     (iii) the effect on radar detection of the sea state, weather and other sources of interference;

     (iv) the possibility that small vessels, ice and other floating objects may not be detected by radar at an adequate range;

     (v) the number, location and movement of vessels detected by radar;

     (vi) the more exact assessment of the visibility that may be possible when radar is used to determine the range of vessels or other objects in the vicinity.

Rule 7

Risk of collision

(a) Every vessel shall use all available means appropriate to the prevailing circumstances and conditions to determine if risk of collision exists. If there is any doubt, such risk shall be deemed to exist.

(b) Proper use shall be made of radar equipment if fitted and operational, including long-range scanning to obtain early warning of risk of collision and radar plotting or equivalent systematic observation of detected objects.

(c) Assumptions shall not be made on the basis of scanty information, especially scanty radar information.

(d) In determining if risk of collision exists the following considerations shall be among those taken into account:

     (i) such risk shall be deemed to exist if the compass bearing of an approaching vessel does not appreciably change;

     (ii) such risk may sometimes exist even when an appreciable bearing change is evident, particularly when approaching a very large vessel or a tow or when approaching a vessel at close range.

Rule 8

Action to avoid collision

(a) Any action taken to avoid collision shall, if the circumstances of the case admit, be positive, made in ample time and with due regard to the observance of good seamanship.

(b) Any alteration of course and/or speed to avoid collision shall, if the circumstances of the case admit, be large enough to be readily apparent to another vessel observing visually or by radar; a succession of small alterations of course and/or speed should be avoided.

(c) If there is sufficient sea room, alteration of course alone may be the most effective action to avoid a close-quarters situation provided that it is made in good time, is substantial and does not result in another close-quarters situation.

(d) Action taken to avoid collision with another vessel shall be such as to result in passing at a safe distance. The effectiveness of the action shall be carefully checked until the other vessel is finally past and clear.

(e) If necessary to avoid collision or allow more time to assess the situation, a vessel shall slacken her speed or take all way off by stopping or reversing her means of propulsion.

Rule 19

Conduct of vessels in restricted visibility

(a) This Rule applies to vessels not in sight of one another when navigating in or near an area of restricted visibility.

(b) Every vessel shall proceed at a safe speed adapted to the prevailing circumstances and conditions of restricted visibility. A power-driven vessel shall have her engines ready for immediate manoeuvre.

(c) Every vessel shall have due regard to the prevailing circumstances and conditions of restricted visibility when complying with the Rules of Section I of this Part.

(d) A vessel which detects by radar alone the presence of another vessel shall determine if a close-quarters situation is developing and/or risk of collision exists. If so, she shall take avoiding action in ample time, provided that when such action consists of an alteration of course, so far as possible the following shall be avoided:

     (i) an alteration of course to port for a vessel forward of the beam, other than for a vessel being overtaken;

     (ii) an alteration of course towards a vessel abeam or abaft the beam.

(e) Except where it has been determined that a risk of collision does not exist, every vessel which hears apparently forward of her beam the fog signal of another vessel, or which cannot avoid a close-quarters situation with another vessel forward of her beam, shall reduce her speed to the minimum at which she can be kept on her course. She shall if necessary take all her way off and in any event navigate with extreme caution until danger of collision is over.

Index

131